World Heritage Sites and Tourism

Not all World Heritage Sites have people living within or close by their boundaries, but many do. The designation of World Heritage status brings a new dimension to the functioning of local communities and particularly through tourism. Too many tourists, accentuated by the World Heritage label, or in some cases not enough tourists, despite anticipation of increased numbers, can act to disrupt and disturb relations within a community and between communities. Either way, tourism can be seen as a form of activity that can generate interest and concern as it is played out within World Heritage Sites. But the relationships that World Heritage Sites and their consequent tourism share with communities are not just a function of the number of tourists. The relationships are complex and ever changing as the communities themselves change and are built upon long-standing and wider contextual factors that stretch beyond tourism.

This volume, drawing upon a wide range of international cases relating to some 33 World Heritage Sites, reveals the multiple dimensions of the relations that exist between the sites and local communities. The designation of the sites can create, obscure and heighten power relations between different parts of a community, between different communities and between the tourism and the heritage sector. Increasingly, the management of World Heritage is not only about the management of buildings and landscapes but also about managing the communities that live and work in or near them.

Laurent Bourdeau is in the Department of Geography at the Université Laval, Canada.

Maria Gravari-Barbas is at the Institut de Recherche et d'Études Supérieures du Tourisme, University of Paris 1 Panthéon-Sorbonne, France.

Mike Robinson is at the Ironbridge International Institute for Cultural Heritage, University of Birmingham, UK.

Heritage, Culture and Identity

Series editor: Brian Graham, University of Ulster, UK

This series explores all notions of heritage – including social and cultural heritage, the meanings of place and identity, multiculturalism, management and planning, tourism, conservation, and the built environment – at all scales from the global to the local. Although primarily geographical in orientation, it is open to other disciplines such as anthropology, history, cultural studies, planning, tourism, architecture/conservation, and local governance and cultural economics.

For a full list of titles in this series, please visit www.routledge.com/Heritage-Culture-and-Identity/book-series/ASHSER-1231

Building a World Heritage City
Sanaa, Yemen
Michele Lamprakos

World Heritage, Tourism and Identity
Inscription and Co-production
Laurent Bourdeau and Maria Gravari-Barbas

Contested Memoryscapes
The Politics of Second World War Commemoration in Singapore
Hamzah Muzaini and Brenda S.A. Yeoh

Cultures of Race and Ethnicity at the Museum
Exploring Post-imperial Heritage Practices
Divya P. Tolia-Kelly

Defining National Heritage
The National Trust from Open Spaces to Popular Culture
Leslie G. Cintron

World Heritage Sites and Tourism
Global and Local Relations
Edited by Laurent Bourdeau, Maria Gravari-Barbas and Mike Robinson

World Heritage Sites and Tourism

Global and Local Relations

Edited by Laurent Bourdeau, Maria Gravari-Barbas and Mike Robinson

Routledge
Taylor & Francis Group
LONDON AND NEW YORK

First published 2017 by Routledge

2 Park Square, Milton Park, Abingdon, Oxfordshire OX14 4RN
711 Third Avenue, New York, NY 10017

Routledge is an imprint of the Taylor & Francis Group, an informa business

First issued in paperback 2018

British Library Cataloguing in Publication Data
A catalogue record for this book is available from the British Library

Library of Congress Cataloging in Publication Data
Names: Bourdeau, Laurent, editor. | Gravari-Barbas, Maria, editor. | Robinson, Mike, 1960– editor.
Title: World heritage sites and tourism : global and local relations / edited by Laurent Bourdeau, Maria Gravari-Barbas and Mike Robinson.
Description: New York, NY : Routledge, 2017. | Series: Heritage, culture, and identity | Includes bibliographical references and index.
Identifiers: LCCN 2016021681 | ISBN 9781409470618 (hardback) | ISBN 9781315546322 (ebook)
Subjects: LCSH: World Heritage areas—Management. | World Heritage areas—Conservation and restoration. | Tourism—Management. | Culture and tourism.
Classification: LCC G140.5 .W6775 | DDC 363.6/9—dc23
LC record available at https://lccn.loc.gov/2016021681

ISBN: 978-1-4094-7061-8 (hbk)
ISBN: 978-1-138-33937-8 (pbk)

Typeset in Times New Roman
by Keystroke, Neville Lodge, Tettenhall, Wolverhampton

Contents

Figures

Tables

Contributors

Alexandra Arellano is an Associate Professor in the School of Human Kinetics at the University of Ottawa, Canada. She completed her PhD in sociology at the University of Lancaster in the United Kingdom. Her research interests in the area of tourism focus on Inca heritage, social inclusion, and the reproduction of poverty, particularly in the south Andean region of Peru.

Michal J. Bardecki received degrees from the University of Guelph and York University, Canada. He holds the position of Professor in the Department of Geography and Environmental Studies and the interdisciplinary graduate programmes in Environmental Applied Science and Management at Ryerson University, Canada, and is on the Executive of the Ryerson Urban Water Centre. He has a continuing research programme investigating the social and economic impacts of tourism in Nepal's protected areas. His recent areas of research also include environmental interpretation, media and environment, sustainable fashion, and environmental assessment.

Bernhard Bauer holds a PhD in cultural anthropology from the University of Vienna, Austria (2014) and has extensive experience in research and consultancy for sustainable tourism, socio-economic, and cultural development. He implements projects and conducts training courses for international development institutions in Africa, South America, Asia, and Europe. His main interest is in urban anthropology, cultural economy, and the sustainable commodification of cultural products for tourism. For his doctoral dissertation Bernhard analysed the commodification of living cultural heritage in New Orleans, USA. In 2015 Bernhard and his wife published their first documentary film about social policies and inequalities in Brazil.

Laurent Bourdeau is Full Professor in the Department of Geography at Université Laval, Québec City, Canada. He is also the holder of the Research Chair on Attractiveness and Innovation in Tourism. His research interests include World Heritage label and consumer values. Professor Bourdeau has seen his works published in many journals, including *Environment and Behavior*, *Journal of Retailing and Consumer Services*, *Museum Management*, *Journal of International Art Management*, and *Journal of Cultural Heritage Management and Sustainable Development*.

Patrick Brouder is a Senior Research Fellow at the School of Tourism and Hospitality, University of Johannesburg, South Africa, and a Research Associate at the Department of Tourism Studies and Geography, Mid Sweden University. He serves on the Steering Committee of the International Polar Tourism Research Network and has been conducting research in Arctic Sweden since 2010.

Ronda L. Brulotte is Associate Professor of Anthropology and Associate Director for Academic Programs at the Latin American and Iberian Institute at the University of New Mexico. She is author of *Between Art and Artifact: Archaeological Replicas and Cultural Production in Oaxaca, Mexico* (University of Texas 2012) and co-editor with Michael A. Di Giovine of *Edible Identities: Food as Cultural Heritage* (Ashgate 2014). Her current work explores the globalization of Oaxacan mezcal as well as the development of northern Mexico's sotol industry.

Alfredo Conti is an architect, specialized in heritage conservation. He acts as ICOMOS expert for the evaluation and monitoring of World Heritage properties and as UNESCO consultant on projects on cultural heritage. He directs the Research Centre on the Territory and the Environment at the Commission for Scientific Research of the province of Buenos Aires and the Institute of Research on Tourism at La Plata University. He is Vice-President of ICOMOS, Professor at La Plata University, and Academic Director of the Postgraduate Course on Heritage and Sustainable Tourism, UNESCO Chair on Cultural Tourism, Buenos Aires.

Jennifer Michelle Cook has a Master's in Applied Science and Environmental Management from Ryerson University, Canada, where she completed her thesis work on 'Valuing Protected Areas: A Contingent Valuation of Tourism in Chitwan National Park, Nepal'. Michelle is a professional project manager supporting local government and first nation community development and planning. Michelle has experience in social impact assessment, environmental assessment, public consultation and engagement, water policy development, and agricultural pest management. Michelle is committed to working toward the better management of local resources for her community. Michelle lives in Kelowna, BC.

Amy Cox Hall is a Visiting Assistant Professor of Anthropology at Amherst College, USA. She holds a PhD in cultural anthropology from the University of Florida and a Master of Science in the history of art from the University of Edinburgh, UK. Her research interests include science and technology studies, photography, heritage, travel, tourism, and the anthropology of history. Her work has appeared in *Ethnohistory*, *History of Photography*, and various edited book collections. She is currently completing a book on the role of photography on the Yale Peruvian Expeditions to Machu Picchu (1911, 1912, 1914–15).

Michael A. Di Giovine is Assistant Professor of Anthropology at West Chester University, USA and an Honorary Fellow at the University of Wisconsin-Madison.

The author of *The Heritage-scape: UNESCO, World Heritage and Tourism* (2009) and *Tourism: Anthropological Insights* (forthcoming), Michael is the co-editor of *Tourism and the Power of Otherness* (2014), *Edible Identities: Food as Cultural Heritage* (2014), *The Seductions of Pilgrimage* (2015), and Lexington's book series *The Anthropology of Tourism: Heritage, Mobility and Society*. Michael is the Convenor of the American Anthropological Association's Anthropology of Tourism Interest Group and a member of the AAA's Task Force on Cultural Heritage.

Carolina Espinoza Camus, Doctor of Science in Education, is an Associate Professor in the School of Anthropology of the National University of Trujillo, Peru. She worked for 14 years as a member of a multidisciplinary team of scientists sponsored by the Museum of Denmark in the Amazonian cloud forest. Recently she supported another research team from the University of Padua. Her research interests are tourism and heritage, and sustainable development focusing on health and nutrition in remote areas of the Andes and the Amazon basin.

Linda Joyce Forristal is Associate Teaching Professor in the Center for Hospitality and Sport Management at Drexel University in Philadelphia, USA. She specializes in destination management with two major streams of research: cultural heritage and natural resources. She believes that destinations can distinguish themselves through their unique culture and heritage, especially historical food ways. She is a long-time member of Les Dames d'Escoffier (LDEI), the International Council on Hotel, Restaurant, and Institutional Education (ICHRIE), and the International Society of Travel and Tourism Educators (ISTTE). She is an online advocate of cultural heritage tourism and nutrient-dense food through her website, www.motherlindas.com.

Ina Freeman, PhD, is currently working as an Associate Professor in the Economics, Business, and Accounting Department, Rockford University, USA. She teaches in marketing and management to undergraduate and graduate students. As a Canadian, she has lived and taught in a number of countries, giving her a unique multicultural perspective on global affairs.

Rachel F. Giraudo is Assistant Professor of Anthropology at the California State University, Northridge, USA. Her research interests include indigeneity and identity politics in southern Africa, cultural heritage, tourism, and development. She is also developing research projects on cannabis legalization and cannabis tourism in the USA. Rachel is a former Board Member of the Anthropology of Tourism Interest Group of the American Anthropological Association (2013–2016) and Associate of the Intellectual Property Issues in Cultural Heritage (IPinCH) project (2011–2016).

Maria Gravari-Barbas is Professor of Geography at the University of Paris 1 Panthéon-Sorbonne, France. She is Director of the IREST (Institute of Research and Higher Studies on Tourism) and EIREST (Interdisciplinary

Research Group on Tourism Studies) of Paris 1 Panthéon-Sorbonne University. She leads the UNESCO Chair 'Culture, Tourism, Development' and coordinates the UNESCO UNITWIN network of the same name. Her research interests include heritage and heritagization mainly in urban and metropolitan contexts.

Stephanie J. Litka is Lecturer in Anthropology at the University of Dayton in Ohio, USA. The author of various presentations and publications on tourism, language contact, and kinship among the Yucatec Maya in southern Mexico, she recently presented a paper on the role of Maya heroes and public memory at the Society for Applied Anthropology annual meeting in Vancouver. She offers courses on tourism studies as part of summer study abroad programmes and plans to collaborate on community-based tourism projects in the town of Coba, Quintana Roo, Mexico

Jennifer C. Lutton's research examines ways in which cultural activists and historic preservationists in the Caribbean mobilize heritage to engage with as well as confront globalizing processes. She has conducted anthropological fieldwork in St. Lucia on heritage practices and tourism. Her research interests include heritage discourses, tourism, creolization, migration, diaspora, and identity. She is an Adjunct Lecturer in Anthropology at the City College of New York–CUNY and teaches courses on race and racism, tourism, and the Caribbean diaspora in New York City.

David McGlade has worked for the past 20 years on Hadrian's Wall Path National Trail where he attempted to raise the general awareness of the vulnerability of archaeological earthworks to visitor and other pressures. In 1996 he established a fixed-point photographic monitoring system to capture, three times a year, the condition of the monument and National Trail; the data set is today the most complete, globally, for any World Heritage Site. In May 2014 he became chairman of the Offa's Dyke Association (ODA).

Mike Robinson is Professor of Cultural Heritage and holds the Chair of Cultural Heritage at the University of Birmingham, UK. He is Director of the Ironbridge International Institute for Cultural Heritage, a long-standing collaboration with the Ironbridge Gorge World Heritage Site. He is founder and editor-in-chief of the *Journal of Tourism and Cultural Change*. For the past 25 years Mike's work has focused upon the relations between heritage, culture, and tourism, and he has published numerous books, articles, and chapters on the various ways in which these realms collide. He works closely with UNESCO and has written reports on tourism and culture and on cultural diversity. He is one of the founding members of the UNESCO/UNITWIN Network on Tourism, Culture and Development, and an expert adviser to the UK National Commission for UNESCO and to the World Heritage Centre's Sustainable Tourism Programme. Mike has undertaken work on heritage and tourism projects in over 30 countries.

James Rollefson is now conducting research in Peru after a career with the National Research Council of Canada (NRC) focused mainly on assessment of projects in renewable energy, environment, and the practical industrial application of sustainable development principles. He is applying his multidisciplinary background to issues such as sustainable agriculture practices in the Amazon cloud forest and the promotion of the interests of vulnerable groups like the porters on the Inca Trail.

Nitin Sinha is a Conservation Architect and Arbitrator, with experience in building conservation projects. He has actively participated in over 30 conservation-related assignment projects, working with both public bodies (central and local governments, municipalities, international donors, various public corporations) and the private sector (NGOs, consulting firms, corporations and associations, etc.). Nitin got his MA degree in architecture from the London Met Business School and is founder of HCIC – Heritage Conservation Initiative Consultants – based in India.

He Sun is an instructor in the Department of Hotel Management at Shandong College of Tourism and Hospitality, China. Previously, she was a visiting scholar at La Rochelle Business School, France. She holds an MS in Tourism Management from Shandong University, China, and is currently a PhD candidate in Business Management at Shandong University. He Sun's research interests are in tourism management, tourism development, and marketing.

Michele Trimarchi, PhD, teaches Economic Analysis of Law (Catanzaro) and Cultural Economics (Bologna). A member of the editorial board of *Creative Industries Journal*, he is founder and president of Tools for Culture.

Evan R. Ward is Associate Professor of History at Brigham Young University, USA, where he teaches courses on the History of Travel and Tourism, Latin American History, and Inter-American Relations. He received his MA and PhD from the University of Georgia. He is the author of *Packaged Vacations: Tourism Development in the Spanish Caribbean* (Gainesville: University Press of Florida, 2008). He is currently writing a book on companionate travel and coming-of-age travel narratives. He resides in Salem, Utah.

Carsten Wergin (Dr Phil) leads the research group 'The Transcultural Heritage of Northwest Australia: Dynamics and Resistances' at Ruprecht-Karls-University Heidelberg, Germany. His academic background is in sociocultural anthropology, media, and transcultural studies, with a wider thematic interest in digital and environmental humanities research, and a regional focus on the Indian Ocean World, drawing on long-term fieldwork in the Mascarene Archipelago and northwest Australia. His first monograph, *Kréol Blouz: Musikalische Inszenierungen von Identität und Kultur* (Böhlau, 2010) is an ethnography of the transcultural music scene of La Réunion (DOM-TOM). Further publications include the edited volume *Musical Performance and the Changing City* (Routledge 2013) and the special journal issue *Materialities of Tourism* (Tourist Studies 2014/3).

Gregor Williams is from Saint Lucia, and graduated in Engineering Science from St. Francis Xavier University in Canada. He has published several articles on history and heritage and edited the book *Unsung War Heroes of Saint Lucia*. He is now a consultant in the areas of heritage, heritage protection, and training. He has served on the boards of five heritage organizations. He was awarded the Doctor of Letters by the University of Sheffield, United Kingdom, in February 2009 in recognition of his work in education, history, and heritage conservation.

Vincenzo Zappino is an economist with international experience in sustainable tourism and economic development. He graduated from the University of Bologna, Italy with a Master's degree in 1996. He is founding partner of the Italian tourism consultancy firm Target Euro and co-creator and CMO of www.toumake.biz, a tourism destination management platform. As a Tourism Development Consultant, Vincenzo has actively participated in more than 60 development projects, focusing his activities on the development and strengthening of micro, small and medium enterprise tourism networks across the world. He has worked with national and local governments, private stakeholders, and multilateral organizations (World Bank, Inter-American Development Bank, United Nations, European Commission).

1 Tourism at World Heritage Sites

Community ambivalence

Maria Gravari-Barbas, Mike Robinson and Laurent Bourdeau

Introduction

Inscription on the World Heritage List brings with it considerable changes relating to the meaning, organization and management of a heritage site. This book deals with the evolutions that take place *before*, *during* and *after* inscription, even several years later and not only concerning the site *per se* but also within the wider socio-cultural context in which it operates. It analyses the complex and interconnected system of players who both orchestrate and experience the changes that emerge with World Heritage designation and examines the supportive or conflictual processes that take place because of inscription on the World Heritage List.

We understand World Heritage Sites spatially and conceptually as performative spaces and places upon which the rehearsed scripts of UNESCO designation, but also extant social, economic and political realities, are enacted by a variety of local, regional, national and international actors or stakeholders (Morisset and Dormaels 2011). From expert bodies, governmental and non-governmental agencies and large and small business interests to local communities, all are engaged with the production and management of World Heritage. Increasingly, domestic and international tourists play an important role, arguably both as the most important audiences for World Heritage and as co-producers; their presence – whether desired, tolerated or rejected – is now a central consideration in the inscription process and the ongoing management, interpretation and sustainability of a site. The interactions of all stakeholders working across different scales in the theatre of World Heritage are open to global scrutiny, and local issues are swiftly and easily escalated to the international level where common issues emerge and tourism appears to be central amongst these.

Analysis of the interface between tourism and World Heritage has gained considerable and much needed attention within the academic community over recent years across a wide range of disciplinary perspectives (Leask and Fyall 2000; Bourdeau et al. 2012; Gravari-Barbas and Jacquot 2013; Gravari-Barbas et al. 2015). The tourism–World Heritage relationship is now a consolidated field of study, handled variously between poles where tourism is accorded a role either as a disruptive or destructive phenomenon or as an instrument for development and intercultural engagement. This common polarization draws upon an increasing number of empirical studies from different parts of the world that focus on different stages of the World Heritage production–consumption continuum. Several texts focus on the increased visitation of heritage sites *after* their inscription on the World Heritage List (Gravari-Barbas and Jacquot 2008; Gravari-Barbas and Jacquot 2011). This increase appears not only as a validation of development strategies based on the social fact of World Heritage (Licciardi and Amirtahmasebi 2012) but also as a worrying phenomenon that World Heritages Sites have to deal with. As von Droste (2012: 12) reminds us, 'tourism constitutes, next to energy supply and the arms trade, the largest sector of the global economy'. The comparison is not innocent. If the increased number of visitors is considered as *a priori* positive, dominant discourses tend to draw attention to the negative aspects of tourism development on local communities and on the intrinsic qualities of heritage. Discourses that position tourism as a form of 'invasion' or threat remain prevalent amongst World Heritage experts or consultants dealing with the 'tourism issue' (du Cros 2006, 2008). In counterpoint, 'new' forms of tourism, characterized as 'cultural', 'sustainable' or 'fair' (see for instance: Cousin 2008; Landorf 2009), are supposed to offer a solution to the 'problem' of tourism, based as they are around site management plans that address visitor management and aim at optimizing the visitor impacts. Most of all, they are supposed to replace the focus upon tourism and the problematic collective designation of the 'tourist' (with some conflation between the tourism 'industry' and 'outside' tourism stakeholders suspected of draining positive tourism impacts away from a site) with a focus on local communities. Of course, the community concept is just as slippery as that of tourism and is frequently referred to in the abstract and as a homogeneous group. Nonetheless, over a number of years there has been increasing focalization of professional and research papers on the 'communities issue' (see for instance: Albert et al. 2012; Grimwade and Carter 2000; Hill et al. 2011; Gravari-Barbas and Jacquot, 2013a, 2013b).

This book draws upon the conclusions of previous works but seeks to distance itself from dominant and persistent discourses that cast 'tourism' as some form of predator. Notwithstanding the realities of many cases, it works on an approach that goes beyond the simple identification of tourism and of tourism impacts *on* World Heritage Sites. It considers tourism not only as a complex action in itself but also as a concept which is bound up with wider socio-economic realities, mobilities and political frameworks which reveal local communities' ambitions, concerns and actions to the world. As part of this, the contributions in this volume, in their different ways, situate tourism as a local–global interaction between the meta-level narratives and policies of *world* heritage and the needs, desires and performances of locality. This 'new' positioning of tourism comes from an awareness, including from the UNESCO World Heritage Center, that it is no longer possible to ignore one of the most powerful social, cultural and economic phenomena of late modernity. But it also comes – at least we would like to introduce this idea here – from an inversion of the tourism–heritage relation paradigm where tourism is no longer seen as a simple *epiphenomenon* of heritage (tourism develops because heritage is already there), but rather as *antecedent* of heritage that relates the social production of heritage (the demand for heritage, the need of heritage, the constitution of heritage) to the real, international demand for tourism. Millions of people travel to experience heritage-based places, artefacts and imaginaries which carry meaning for today's world (Gravari-Barbas 2012). So, far from being a marginal element present at the beginnings of the World Heritage Convention, tourism can be recognized today as a phenomenon contributing to the very fulfilment of the constitutive act of UNESCO because it offers the possibility to associate with World Heritage Sites and pluralized, cross-cultural, cross-national gazes and practices.

Wider contexts and new dynamics for heritage

Specific questions relating to tourism and World Heritage Sites are framed by several broader trends in contemporary society. The continuous production of 'new' and diverse heritage points to a process of questioning of hierarchies and increasing democratization whereby the structures and meanings of heritage as grand institutions, largely inherited from nineteenth century European patrimony and nation building, are changing. The tendency towards a de-institutionalization of heritage, greater levels of local participation, strongly articulated community voices and social action, aided by open access to global media, all combine to allow for increased levels of the 'bottom-up' production of heritage. This can be a somewhat messy, almost anarchistic, process (Robinson 2011) and is complex and fraught with contradictions (Gravari-Barbas and Jacquot 2013b). The state still has a critical and formal role to play in heritage production, but what constitutes heritage can now be seen as part of a social contract, of a certain cooperative reality, which is shifting attention from heritage as object towards a *relationship* established between object and a diverse and shifting audience. A further part of this re-fashioning of the heritage–audience relationship is the de-mystification/de-consecration of heritage that closes the distance between it and social groups and allows for wider and multiple readings that can then generate different meanings. This is greatly amplified through the availability of digital mediation tools which can generate a sense of potential; a more individualized view of heritage; the chance to integrate fact and fiction; a removing of the taboo around some heritage objects, as they are stripped of the decorum of their institution, removed from the realm of contemplation and thrust into the world of daily usage; and a shift in the position of experts and their areas of interest (Robinson and Silverman 2015).

The policy and practice of heritage conservation is marked by an evolution of a logical dynamic. Structures and systems for the scheduling and regulation of heritage have themselves been inherited and have in part provided some stability for the protection of what we may term a 'traditional' heritage. But these systems continue to be challenged and destabilized by a number of factors, such as the rapid pace of urban development, global mobilities, the growing importance of the tourism industry, as well as changing sensibilities aligned with the production of new and alternative heritages, all of which are contributing to the remaking of old traditions, the invention of new ones and a new order of authenticity. The changes mark a shift from a 'stock' heritage towards a 'flux' heritage which acts as an *active principle of sustainable development* linking lifestyle and sustainable management of local resources in a more organic way. The notion of the 'cultural landscape', the bridging of tangible, intangible and living heritage elements (Rössler 2006; Taylor and Lennon 2010), together with notions of serial sites and trans-boundary heritage, all express a reorganization of the world and the need for the heritage sector to adapt.

The sustainability model in particular asserts a certain way of reading the world and the relationships that individuals have with the world. It is influencing how heritage is approached and is challenging the long-standing predominance of aesthetics by infusing the discourse with ethical concerns regarding issues such as the reuse or recycling of heritage sites and objects for social use or social good and the rights that communities and individuals have with regard to accessing their own heritage. In the work of directing heritage, both through UNESCO and the Council of Europe, we can identify shifts towards a sustainable approach to heritage that recognizes the centrality of communities in its utilization and its management and is manifest in the Budapest Declaration on World Heritage (World Heritage Committee 2002) and the Council of Europe's (2005) Framework Convention on the Value of Cultural Heritage for Society – the 'Faro Convention'.

Reappraising community engagement with heritage

Stemming from a World Heritage Committee meeting in Christchurch in 2007, a proposal from the New Zealand State Party resulted in the adoption of 'Community' as the fifth (fifth 'C') of the World Heritage Convention's Strategic Objectives.[1] This additional objective confirmed that, 'in the future, the conservation of the world's natural and cultural heritage should, wherever possible, be done with the active engagement of communities which have a close relationship with the heritage in question'. Furthermore, it pledged that State Parties, 'as appropriate, seek the active involvement of communities at all stages, from the preparation of tentative lists through to conservation requirements for sites which are in danger' (World Heritage Committee 2007).

While it can be argued that recognition of the role of communities as an intimate and intrinsic part of the World Heritage concept was slow to materialize, in the policy discourse that surrounds World Heritage today the 'fifth C' of the Convention has gained considerable momentum. That said, as with many aspects of the Convention, the onus is upon the State Parties for implementation, and the attention given to the role of the community will vary in accordance with the systems of governance that are operational at state level. Moreover, although recognition of the role of the community in the creation, protection and management of World Heritage was a step forward in policy terms, in practice any consideration of the community moves us from a rather essentialized and abstract concept to a more complex and dynamic social reality that is problematic to identify and work with. In the first instance, as has been widely discussed in the literature (e.g. Cohen 1985; Calhoun 1991; Delanty 2003), communities are not homogeneous or always animated by common goals and objectives. Rather, they are frequently fragmented and marked by fluctuating asymmetries of power. Moreover, while we may often refer to communities as 'local', the scale of a community as a stakeholder will vary.

This book brings together cases and examples relating to 33 different World Heritage Sites located in four continents (see Table 1.1). At its heart the book examines the implications of inscription on the World Heritage List from the local level, but it also addresses the repercussions at the global level. Tourism is examined as an important mediator, as a *factor of change*, as a *catalyst*, and as a *guiding principle* of the decisions taken (implicitly and explicitly) both locally and globally. As the book's subtitle – 'global and local relations' – implies, there is a critical dialectic between guidelines and imperatives set at the global level and how these are understood and interpreted at the local scale (Bianchi 2013). In this 'glocal' framework for heritage (Salazar 2005), the community issue is central.

Table 1.1 World Heritage case studies examined in this volume

World Heritage Site	*Date of inscription*	*Country*	*Continent*
Tsodilo	2001	Botswana	Africa
Chitwan National Park	1984	Nepal	Asia
Group of monuments at Hampi	1986	India	
Mount Taishan	1987	China	
Ensemble of monuments of Huế; sanctuary of Mỹ Sơn; old city of Hội An	1993; 1999; 1999	Vietnam	
Temple and cemetery of Confucius and the Kong family mansion in Qufu	1994	China	
Laponia Area	1996	Sweden	Europe
Benevento, Santa Sophia; part of the 'Longobards in Italy: Places of the Power'	2011	Italy	
Historic sanctuary of Machu Picchu	1983	Peru	South America
City of Cusco	1983	Peru	
Historic quarter of the city of Colonia del Sacramento	1995	Uruguay	
Historic quarter of the seaport city of Valparaíso	2003	Chile	
Pitons Management Area	2004	Santa Lucia	
Historic centre of Oaxaca and archaeological site of Monte Albán	1987	Mexico	Central America
Pre-Hispanic city of Chichén Itzá	1988	Mexico	

(continued)

Table 1.1 Continued

World Heritage Site	*Date of inscription*	*Country*	*Continent*
Town of San Miguel and the sanctuary of Jesús Nazareno de Atotonilco	2008	Mexico	
Yellowstone	1978	USA, Wyoming	North America
Mesa Verde	1978	Colorado	
Grand Canyon	1979	Arizona	
Everglades	1979	Florida	
Independence NHS	1979	Pennsylvania	
Wrangell-St. Elias	1979	Alaska	
Redwood	1980	California	
Mammoth Cave	1981	Kentucky	
Olympic	1981	Washington	
Great Smoky Mountains	1983	Tennessee/North Carolina	
San Juan NHS	1983	Puerto Rico	
Yosemite	1984	USA, California	
Statue of Liberty	1984	New York	
Hawaii Volcanoes	1987	Hawaii	
Chaco culture	1987	New Mexico	
Glacier (Waterton)	1995	Montana	
Carlsbad Caverns	1995	New Mexico	

Also implicit throughout the cases is the question of who benefits from tourism development at World Heritage Sites? Again the community issue is central. Various assumptions are made relating to the extent to which local communities benefit (and indeed dis-benefit) from tourism brought about through development associated with UNESCO listing. At one level this is an exercise in measurement – of the development and economic impact. But on another level it is a political issue which raises questions about who is involved in the processes of identification, preservation and valuation of World Heritage properties. Depending on the system of governance in place, the role of the community varies. Very often the process of designating World Heritage has been undertaken by experts, bypassing community knowledge and voices and resulting in a politics of exclusion. But what is the place of 'non-experts' in World Heritage-making, and the associated tourism development, and how does capacity building (Harrison 2013) materialize so as to shape the strategies and define the policies and projects that build on the act of UNESCO inscription?

From displacement to empowerment: communities at the centre

A number of researchers have sought to evidence the negative role that World Heritage and related tourism can play in the apparent destruction of local community traditions and wider social life (Burns 1999; Hampton 2005; Evans 2002). In the case of the World Heritage Sites Borobudur and Prambanan in Indonesia, Wall and Black (2004) argue that the values that local people attach to heritage are often not adequately represented or respected by other participants in the planning and management of sites, to the detriment of both the plans and the people. The authors underline that the tendency to adopt 'top-down' planning procedures has resulted in the disenfranchisement of local people and has tended to freeze sites and displace human activities, effectively excluding local people from their own heritage. Evidence of negative impacts on local communities at other World Heritage Sites, such as Luxor (Gamblin 2007), Tlakotalpan in Mexico (Alavez 2013) or the Pitons Management Area (Nicholas et al. 2009), has been well discussed and in part has helped to shape a debate within UNESCO around the broad theme of 'communities'. Following the addition of 'Community' as a fifth Strategic Objective to the four 'Cs' of the Budapest Declaration (World Heritage Committee 2002), the idea was to encourage the participation of communities in both identifying and managing World Heritage. The creation of the 'cultural landscape' category had already moved the heritage concept forward by refocusing on local conceptions of heritage (Sullivan 2004), thus reconfiguring the relationship between nature and culture after the great division generated by modernity (Latour 1991). Indeed, the nominations of the first cultural landscapes put the involvement of local people at the very centre of evaluations, inscriptions and management practices; at Uluru-Kata Tjuta National Park (Australia), the reclassification of the site as a cultural landscape changed its management plan, allowing Aboriginal people, the owners of the land, to be part of the site management (Rössler, 2012).[2]

However, this integration of communities concerns not only heritage recognition and the pluralization of the meaning of 'outstanding universal value' (Cameron 2005; Djament-Tran et al. 2012). The community focused resolution adopted in Christchurch also invited State Parties to recognize the capacity of communities and the relevance of their contributions to the definition of the conditions of conservation and heritage enhancement. Thus, communities are not just the object but also the subject of development policies and sustainable tourism articulated in the management of UNESCO sites. The appropriation of sustainable tourism by the communities refers to their direct participation in tourist activity and to the dynamics of empowerment, allowing them representation and asserting themselves as actors in the development process (Cole 2007; Baldwin and Chandler 2010). Thus, in Colombia, in the Archaeological Park of San Augustin, inscribed on the World Heritage List in 1995, the Yanacona community is constituted within the wider indigenous movement that opposes restrictions on development in order to ensure the conservation of the Park. This community has appropriated the strategic discourse developed by UNESCO on the role of communities in the management of sites and claims the right to development, including the development of transportation infrastructure and the creation of a market for the sale of indigenous crafts to visitors (Jackson and Ramírez 2009). This example demonstrates the non-consensual dimension of active participation which incorporates community discourse and which may come into opposition with public institutions (Langfield 2010).

Rössler (2012) develops the example of the Kayas, a series of eleven forest areas along 200 kilometres of the Kenyan coast. This property is only in existence because the local communities (the Mijikenda) have protected these forests. The property, inscribed on the World Heritage List in July 2008, takes into consideration the fact that these forests are a fundamental source of the Mijikenda's sense of 'being-in-the-world' and of place. As Rössler (2012: 28) underlines:

> The statement demonstrates clearly the long evolution since the 1992 decision to include cultural landscapes on the World Heritage List and the 1995 change to the Operational Guidelines to include people in the nomination process. It illustrates the fundamental recognition of customary law and traditional management practice by the living communities.

The traditions and management practices of communities recognized as being fundamental to the being of some World Heritage Sites, with a necessity to maintain the conditions of these practices, can, however, instil in us a rather romantic and reified vision of harmonious communities. They are not stable, homogeneous groups but are frequently characterized by divisions with mixed attitudes and interests. Furthermore, they have their own dynamics, with changes being brought about by normative socio-demographic change and shifting power relations. Threats to community identity and/or cultural resources can function to bind communities together, while at the same time they can act to generate self-interrogation of their social structure. The recognition and development of local heritage and its escalation to World Heritage status can both build a community (Waterton and Smith 2010) and divide it. Either way, their integration in tourism policies and sustainable development related to World Heritage positions communities as important players in the construction and management of sites, ranging from participation in economic activity (Su and Wall 2012) and the mediation of sites to the development of policy and political decision making.

Communities are taken to be the target group to benefit from the wider agenda of sustainable development activated through tourism at World Heritage Sites. This was the theme of the 40th anniversary of the World Heritage Convention in 2012 and is manifest in the World Heritage Sustainable Tourism Programme launched at the World Heritage Anniversary Meeting in Kyoto in the same year. The Programme sees communities not only as central to the conservation and management of sites but through capacity building also as recipients of economic benefits that can flow from sustainable tourism integrated with other aspects of the management of the site (Istasse 2011). Galla (2012), in a specially commissioned anniversary book of case studies, *Benefits beyond Borders*, exemplifies this with examples of how local communities can harness the tourism potential from World Heritage – in terms of both the physical site and the UNESCO World Heritage 'brand' – to achieve development goals.

The explicit connections between World Heritage status and tourism-related development through economic and social betterment can be seen as a lever for further community engagement. World Heritage becomes not a goal in itself but more of a mechanism that communities can harness for development. This in turn has raised the stakes regarding who in the community does benefit and how, where and when. Issues of territorial rights and ownership have been accentuated or re-ignited in recognition of the latent and actual benefits that can flow from site inscription. Mumma (2003), for instance, discusses in the context of World Heritage the re-instatement of the legitimacy of community-based legal systems, which involves re-instating historic ownership and/or historic rights for using resources, particularly land.

The reconfiguration of the community in World Heritage terms can be viewed as part of a more sweeping shift in emphasis within heritage policy. The Council of Europe's Faro Framework Convention on the Value of Cultural Heritage for Society (2005) has contributed considerably to de-centring the focus from heritage to the relationship that communities build with heritage; from the right *of* heritage to the communities' right *for* heritage. Arguably, in terms of UNESCO discourse and the boundaries of the World Heritage Convention, the focus is still grounded in the outstanding universal value (OUV) of the property and the needs of the associated communities remain somewhat distant from that critical criterion. The OUV is recognition of significance to an imagined 'global' community and can (and has) bypass(ed) the needs or desires of the actual local community so that engagement with the latter is very much a *post hoc* intervention. The extent to which communities want to engage in the World Heritage process at all remains a critical question and invokes consideration of the wider systems of governance for the State Parties (Robinson 2000).

The process of designating World Heritage has moved forward, and continues to do so, toward a more participatory system where communities can – potentially – share the benefits that inscription can bring, but as yet mechanisms to fully involve local communities in the core decision making remain limited and variable. But of course this falls within a definition of 'community' that is marked by its territorial remit as 'local', implying not just proximity to a site but other attributes such as continuity, fixidity and a sense of identity linked to the meaning of a site. However, there are other communities engaged with World Heritage. Discussions held in Christchurch in 2007 on communities reiterated the distinction between the close and the distant. In the proposal document that went to the World Heritage Committee at Christchurch (WHC-07/31.COM/13B, p. 1) the point was made that the term 'communities' involves all forms of non-state actors:

> That is, from the smallest groups of citizens, in whichever form they manifest themselves. They may range from groupings of peoples as indigenous, traditional and/or local peoples. They may be presented as, inter alia, community groups, tribes, non-governmental organizations, private enterprise and/or local authorities. The defining characteristic of communities, in this setting, is what they possess. They all possess a direct connection, with relevant

interests, to individual sites and often they have a connection that has endured over time. Typically, these communities share a close proximity with the sites in question. These peoples and/or entities are not necessarily directly representing official State positions, and may actually be in dissent from official positions.

The World Heritage concept works on the basis of a global community idea with emotional attachments capable of being formed with sites at a distance. Concern for the protection of World Heritage transcends the notion of the immediate community. In the context of tourism, this attachment is direct through tourists and, although temporary, their role can be influential. Visitors, in the broad sense of the term, are regularly mentioned as actors in the conservation and valuation of World Heritage through sustainable tourism. This is translated into a whole range of devices aimed at spreading awareness about the outstanding universal value and a commitment to World Heritage, such as an agreement that was implemented between the World Heritage Centre and Tripadvisor which allowed visitors to communicate their attachments to and concerns about their experiences at World Heritage Sites effectively transformed tourists into a virtual community keeping a vigil on heritage. At the same time, the involvement of visitors through this monitoring process provided legitimacy to various subjectivities, producing a kaleidoscopic vision of World Heritage (Gravari-Barbas and Jacquot 2013b). Global tourist engagement as a community of interest can also be seen developing through 'crowd-funding' projects that allow transnational engagement with World Heritage conservation projects.

Contributions to this book

In this book we work with a wide interpretation of the term community, taking into consideration various attachments to World Heritage at various space scales and for multiple reasons. For some communities, World Heritage is a very localized phenomenon, and considerations of relationships to it are condensed into practical daily life. For other communities the World Heritage concept opens them up to the global arena/global marketplace and is producing new connectivities with communities elsewhere through tourism.

Utilizing data from Việtnam and Italy, Michael A. Di Giovine (Chapter 2) argues that social 'betterment' and local empowerment are produced not (merely) through economic influxes, but rather through the valorization of a group's worldview, which both UNESCO's designation and foreign attention (in the form of tourism) can bestow upon a place of local interest. Yet in reconceptualizing the vital role that communities play in both site management and touristic interactions with a World Heritage Site, Di Giovine contends that adopting this revitalization paradigm can better preserve the 'spirit of place' despite often intense tourism development and potentially harmful re-presentations of the site.

In Chapter 3 James Rollefson, Carolina Espinoza Camus and Alexandra Arellano examine the inscription of the archaeological complex of Machu Picchu as a World Heritage Site (1983) and how that contributed to transforming the Inca citadel into the most important tourist destination in Peru. They explore tourism development in the region and discuss some of its costs and benefits for the local communities and tourism workers. Their analysis looks at regional inequalities in the Peruvian context and argues that the World Heritage designation seems to work as a global appropriation not only of local heritage but also of the poor who work in the tourism industry. After first outlining tourism in Peru and in Cusco, which boomed in the 1990s, the chapter examines the case of the porters of the Inca Trail and evaluates the impact of their integration in the tourism system and World Heritage route.

The site of Machu Picchu is also examined by Amy Cox Hall in Chapter 4, who concentrates her attention on the ways in which Cusqueño tour guides to Machu Picchu are caught up in the contradictions of various understandings of globality and locality, often framed as a dichotomy between *being Incan* and *becoming cosmopolitan*. While guiding tourists to Machu Picchu is often viewed as a soul-fortifying sojourn to their 'sanctuary', a place where one can connect with their ancestors, the city of Cusco is framed as a historical space caught in a moment of rapid and, at times, dangerous change. At the same time, becoming cosmopolitan is part of being a successful tour guide. Hall examines the ambivalences produced around patrimony and modernity through becoming a UNESCO World Heritage Site – a heritage for all humanity. Ambivalence is also a theme in the case of the Pitons (St. Lucia) examined by Jennifer C. Lutton and Gregor Williams. They look at how the pursuit of neoliberal development strategies which prioritize tourism, marketing and branding of the Pitons creates a tension amongst many St. Lucians regarding the relevance of the World Heritage listing. The universal imperatives of World Heritage become problematic as the tourism industry seeks to cater for a global elite but at the same time acts to restrict the access of St. Lucians to the site.

In Chapter 6 Ronda L. Brulotte discusses the case of Oaxaca, Mexico, declared a UNESCO World Heritage Site in 1987 and today the region's premiere tourist attraction. For approximately 30 years, residents from the nearby town of San Antonio Arrazola have made and sold replicas of pre-Hispanic artifacts to tourists at nearby Monte Albán. Yet in contrast with other Oaxacan crafts, which are promoted by the state as authentic expressions of local culture, the production and sale of replica crafts has been a controversial and often maligned practice. Officials from the National Institute of Anthropology and History (INAH), including archaeologists, have actively tried to remove replica sellers from the site, claiming that they have neither a legitimate historical or cultural connection to Monte Albán nor an interest in its cultural and scientific value. Vendors, on the other hand, have a different understanding of their historical relationship to the archeological zone, and struggle to remain in what they consider to be their legitimate place of work. Since the appearance of the first vendors at the site in the 1960s, clashes between INAH representatives and vendors have been commonplace and at times have escalated to the point of violence and police intervention. Also in Mexico, Stephanie J. Litka (Chapter 7) examines the ways local employees mediate between tourist, government and private businesses' expectations of ideological, monetary and spatial management. She shows how the Maya strategically utilize these tensions to construct a dynamic sense of identity through space and time and argues that contemporary Mayan identity exists as a process that depends on these tensions as opposed to an entity that exists as a mere reaction to tourism.

In Chapter 8 Rachel F. Giraudo shows that, in South Africa, development facilitators view cultural heritage as an economic resource to be utilized for poverty alleviation, but they generally pay less attention to the social impacts of heritage tourism on local communities. Through a heritage ethnography approach, Rachel examines how community-based heritage tourism affects the ethnic and gender identity politics of the Ju|'hoan and Hambukushu communities living near the Tsodilo World Heritage Site in Botswana. She analyses the commodification of their intangible heritage, as they must now resort to a cash economy for self-reliance by means of heritage tourism.

Bernhard Bauer, Nitin Sinha, Michele Trimarchi and Vincenzo Zappino (Chapter 9) show that, for the World Heritage Site of Hampi in India, a participatory development approach representing an empowerment strategy for the local population enhances the cooperation between different actors in the tourism development process through local consultative committees (including locals, governmental institutions, public and private tourism stakeholders, site managers, etc.). In this way the community's role grows more important so stimulating further participation and an interdependency between actors in the tourism sector.

In Chapter 10 Carsten Wergin sketches some of the discursive strategies and multi-layered relationships at work on the island of Rodrigues in the Le Morne cultural landscape in the Indian Ocean. He maps out the dynamic process of simultaneous preservation and development of Rodriguan nature and culture and questions the extent to which local people accept their 'natural' environment as something that needs to be protected under international rules and regulations, and how particular forms of Rodriguan culture can be brought in line with globalist understandings of World Heritage.

Patrick Brouder (Chapter 11) focuses upon the business community and empirically investigates the tourism sector in northern Sweden regarding nature, culture and tourism development. Through a quantitative survey of tourism entrepreneurs he explores how the Laponia World Heritage Area is viewed and valued by tourism businesses in the county of Norrbotten, where Laponia is located, and also how tourism entrepreneurs value nature and culture in general with regard to tourism. Still within the realms of the tourism community, Linda Joyce Forristal in Chapter 12 examines US national parks that are also World Heritage Sites and how themed or named objects in an archived and searchable public record, such as WorldCat, can be a reliable predictor of visitation for the sites. She addresses the practical implications, which include the amount of information created and stored in the public record and how this may have a direct impact on the level of visitation stemming from increased awareness and that impacts destination image and choice, and ultimately influences visitation. She suggests that monitoring and managing the relationship between documents and visitation may be of value to managers, governments and concerned residents in gateway communities associated with World Heritage Sites. Staying with the idea of the business community, Evan R. Ward in Chapter 13 examines the impact that private development, in the form of Panagra Airlines, has played in promoting Machu Picchu prior to its state development with input from UNESCO and the Organization of American States. In Chapter 14 Ina Freeman and He Sun, looking at World Heritage Sites in Shandong Province, China, recognize the natural competition between operators of similar tourism services and the need for these operators to coordinate with other agencies and operators to maximize tourists' experience. Within this cooperation, the role of the government is one of coordination with the tourism firms and the continued promotion of private enterprise through the various involvements of the local communities. In Chapter 15 Jennifer Michelle Cook and Michal J. Bardecki explore the willingness of foreign tourists to pay to support conservation and environmental efforts at Chitwan National Park in Nepal. They confirm that an increase in the entry fee could provide valuable additional resources which could go towards conservation efforts in the park itself and supporting sustainable development initiatives in buffer zone communities.

In Chapter 16 Alfredo Conti analyses the case of several historic towns in Latin America and the Caribbean that are inscribed on the World Heritage List. He draws attention to the fact that lack of proper tourism planning and community engagement causes several kinds of problems regarding authenticity and preservation of the outstanding universal values of these unique or exceptional towns. The tension between a community's desire to benefit economically through tourism and the need to preserve the outstanding universal value of a site is a persistent theme.

David McGlade in Chapter 17 analyses Hadrian's Wall, one of the UK's most iconic landscapes and also one of its most protected. Since the opening in May 2003 of Hadrian's Wall Path National Trail overall visitor numbers to the ancient monument have increased. These additional visitors bring not only economic benefits to the local rural communities but also new visitor management challenges. McGlade asks: what are the limits of acceptable change and who decides?

The cases discussed in this volume cover only part of a complex series of relationships that exist and evolve between World Heritage Sites and their attendant local communities. It is clear that symbiotic relationships are possible prior to and after site inscription. The benefits that well-managed sustainable tourism can bring both to the heritage site and to the community can ease and mitigate the upheaval that some communities may experience in the face of increased visitor numbers. At the same time, a lack of visitors, in spite of the World Heritage label, can stimulate unease and resentment within communities. It is tourism and the tensions it can generate and highlight that can disturb and disrupt communities. This point is increasingly recognized by State Parties and is enshrined in the World Heritage Sustainable Tourism Programme that seeks a balanced approach. Of course each case is different and rooted in local histories, different systems of governance and where communities are situated on the spectrum of economic and social development. For the majority of communities, the designation of World Heritage status comes very late in the relationship that has evolved between a site and the local population. Moreover, the implications of the changes that may or may not manifest themselves are not always fully identified or explored. Good governance and the enactment of local democracy would appear to be prerequisites so as to engage local communities not only before World Heritage status is declared but also from the earliest stages of the inscription process, with sufficient weight being given to the possibility of opting out of the World Heritage system once the consequences have been explored as fully as possible. This is easier said than done for states wishing to be part of the World Heritage network and also as communities themselves change as a function of wider variables. Until then, and for those sites that have populations in or near to their boundaries and are thus impacted by both the positives and the negatives of tourism, community ambivalence may well be the norm.

Notes

1 The other four 'C's are: strengthen the **Credibility** of the World Heritage List, as a representative and geographically balanced testimony of cultural and natural properties of outstanding universal value; ensure the effective **Conservation** of World Heritage properties; promote the development of effective **Capacity-building** measures, ncluding assistance for preparing the nomination of properties for the World Heritage List and for the understanding and implementation of the World Heritage Convention and related instruments; increase public awareness, involvement and support for World Heritage through **Communication**.

2 'We, the traditional land owners of Uluru-Kata Tjuta National Park, are direct descendants of the beings who created our lands during the *Tjukurpa* (Creation Time). We have always been here. We call ourselves *Anangu*, and would like you to use that term for us' (Nellie Patterson, traditional owner, at www.deh.gov.au/parks/uluru/index.html, the official website for Ayer's Rock).

References

Alavez, F. (2013) 'Développement touristique et durabilité dans deux sites du Patrimoine Mondial: Le site préhispanique de Tajin et la ville historique de Tlacotalpan, Mexique", PhD thesis, Paris 1 University-Sorbonne.

Albert, M.T., Richon, M., Viñals, M.J. and Witcomb, A. (eds) (2012) *Community Development through World Heritage*, World Heritage Papers No. 31, Paris: UNESCO.

Baldwin, C. and Chandler, L. (2010) 'At the Water's Edge: Community Voices on Climate Change', *Local Environment: The International Journal of Justice and Sustainability*, 15 (7): 637–649.

Bianchi, R. (2003) 'Place and power in tourism development: tracing the global articulations of community and locality', *Pasos*, 1: 13–32.

Bourdeau, L., Gravari-Barbas, M. and Robinson, M. (eds) (2012) *Tourisme et patrimoine mondial*, Quebec: Presses Universitaires de Laval.

Burns, P. (1999) 'Paradoxes in Planning Tourism: Elitism or Brutalism?', *Annals of Tourism Research*, 26 (2): 329–348.
Calhoun, C. (1991) 'Indirect Relationships and Imagined Communities: Large Scale Social Integration and the Transformation of Everyday Life'. In Bourdieu, Pierre and Coleman, James S. (eds) *Social Theory for a Changing Society*, Boulder, CO: Westview Press, pp. 95–120.
Cameron, C. (2005) 'Evolution of the Application of "Outstanding Universal Value" for Cultural and Natural Heritage', Special Expert Meeting of the World Heritage Convention: The Concept of Outstanding Universal Value; Kazan, Republic of Tatarstan, Russian Federation, WHC-05/29.COM/INF.9B. In Jokilehto, J. (ed.) *The World Heritage List: What Is OUV? Defining the Outstanding Universal Value of Cultural World Heritage Properties*, Berlin: Hendrik Bäßler Verlag, pp. 71–74.
Cohen, A.P. (1985) *The Symbolic Construction of Community*, London: Routledge.
Cole, S. (2007) 'Beyond Authenticity and Commodification', *Annals of Tourism Research*, 34 (4): 943–960.
Council of Europe (2005) Framework Convention on the Value of Cultural Heritage for Society, Faro, 27/10/2005, CETS, No. 199, Strasbourg: Council of Europe.
Cousin, S. (2008) 'L'Unesco et la doctrine du tourisme culturel: Généalogie d'un bon tourisme' *Civilisations: Revue Internationale d'Anthropologie et de Sciences Sociales*, 57, *Tourisme, mobilités et altérités contemporaines*: 41–56.
Delanty, G. (2003) *Community*, London: Routledge.
Djament-Tran, G., Fagnoni, E. and Jacquot, S. (2012) 'La Construction de la Valeur Universelle Exceptionnelle dans la Valorisation des Sites du Patrimoine Mondial, entre Local et Mondial'. In Bourdeau L., Gravari-Barbas M. and Robinson, M. (eds) *Tourisme et Patrimoine Mondial*, Quebec: Presses Universitaires de Laval, pp. 217–234.
Du-Cros, H. (2006) 'Managing Visitor Impacts at Lijiang, China'. In Leask, A. and Fyall, A. (eds) *Managing World Heritage Sites*, Oxford: Butterworth-Heinemann, pp. 205–15.
Du Cros, H. (2008) 'Too Much of a Good Thing? Visitor Congestion Management Issues for Popular World Heritage Tourist Attractions', *Journal of Heritage Tourism*, 2 (3): 225–238.
Evans, G. (2002) 'Living in a World Heritage City: Stakeholders in the Dialectic of the Universal and Particular', *International Journal of Heritage Studies*, 8 (2): 117–135.
Galla, A. (ed.) (2012) *World Heritage: Benefits beyond Borders*, UNESCO Publishing, Cambridge University Press.
Gamblin, S. (2007) 'Tourisme international: État et sociétés locales en Égypte. Louxor, un haut-lieu disputé', PhD dissertation, Institut d' Études Politiques de Paris.
Gravari-Barbas, M. (2012) 'Tourisme et patrimoine, le temps des synergies?' In Khaznadar, C., *Le patrimoine, oui, mais quel patrimoine?* Internationale de l'imaginaire, No. 27, Paris: Babel, pp. 375–399.
Gravari-Barbas, M. and Jacquot, S. (2008) 'Estimation de l'impact de l'inscription des sites du patrimoine mondial.' In Prud'homme, R., Gravari-Barbas, M., Jacquot, S., Talandier, M., Henri-Nicot, B. and Odzirlik, B. (eds), *Les impacts socio-économiques de l'inscription d'un site sur la liste du patrimoine mondial: trois études*, Paris, Centre du Patrimoine Mondial de l'UNESCO, pp. 16–72.
Gravari-Barbas, M. and Jacquot, S. (2011) 'Les impacts de l'inscription au patrimoine mondial de l'humanité: Discours et motivations d'acteurs.' In Furt, J.M. and Fazi, A. (eds) *Vivre du patrimoine, un nouveau modèle de développement*, Paris: L'Harmattan pp. 253–270.

Gravari-Barbas, M. and Jacquot, S. (2013) *Patrimoine mondial et gouvernance des pôles touristiques*, Actes du 3eme séminaire organisée par la Chaire UNESCO Culture, Tourisme, Développement (Université Paris 1 Panthéon Sorbonne), la Convention France – UNESCO, l'Association des biens français patrimoine mondial, ICOMOS.

Gravari-Barbas, M. and Jacquot, S. (2013a) 'Patrimoine mondial, tourisme et développement durable en Afrique.' Special issue of *Via@ Tourism Review*. Online, www.viatourismreview.net/Revue.php (accessed July 2016).

Gravari-Barbas, M. and Jacquot, S. (2013b) *Patrimoine mondial et développement au défi du tourisme durable*. Quebec: Presses de l'Université du Québec.

Gravari-Barbas, M., Bourdeau, L. and Robinson, M. (2015) (eds) *World Heritage, Tourism and Identity: Inscription and Co-production*, London: Ashgate.

Grimwade G. and Carter, B. (2000) 'Managing Small Heritage Sites with Interpretation and Community Involvement', *International Journal of Heritage Studies*, 6 (1): 33–48.

Hampton, M. (2005) 'Heritage, local communities and economic development', *Annals of Toursm Research*, 32 (3): 735–759.

Harrison, R. (2013) *Heritage: Critical Approaches*, London: Routledge.

Hill, R., Cullen-Unsworth, L., Talbot, L. and McIntyre-Tamwoy, S. (2011) 'Empowering Indigenous Peoples' Biocultural Diversity through World Heritage Cultural Landscapes: A Case Study from the Australian Humid Tropical Forests', *International Journal of Heritage Studies*, 17 (6): 571–591.

Istasse, M. (2011) 'Circulation et rencontre du patrimoine et du tourisme dans la médina de Fès: de l'investigation de diverses formes patrimoniales', *Téoros*, 30 (2): 37–46.

Jackson, J.E. and Ramírez, M.C. (2009) 'Traditional, transnational, and cosmopolitan: The Colombian Yanacona look to the past and to the future', *American Ethnologist*, 36: 521–544.

Landorf, C. (2009) 'Managing for Sustainable Tourism: A Review of Six Cultural World Heritage Sites', *Journal of Sustainable Tourism*, 17 (1): 53–70.

Langfield, M. (2010) 'Global Heritage: Perspectives from the Northern Territory, Australia', *International Journal of Heritage Studies*, 16 (3): 187–206.

Latour, B. (1991) *Nous n'avons jamais été modernes: Essai d'anthropologie symétrique*, Paris: La Découverte.

Leask, A. and Fyall, A. (2000) World Heritage Sites: Current Issues and Future Implications', in Robinson, M., Evans, N. and Sharpley, R. (eds) *Tourism and Heritage Relationships: Global, National and Local Perspectives*, Gateshead: Athenaeum Press, pp. 287–300.

Licciardi, G. and Amirtahmasebi, M. (2012) *The Economics of Uniqueness: Investing in Historic City Cores and Cultural Heritage Assets for Sustainable Development*, Washington, DC: World Bank.

Morisset, L.K. and Dormaels, A. (2011) 'Patrimoine mondial: les enjeux locaux', *Téoros*, 30 (2): 3–5.

Mumma, A. (2003) 'Community-Based Legal Systems and the Management of World Heritage Sites', In de Merode, E., Smeets, R. and Westrik, C. (eds) *Linking Universal and Local Values: Managing a Sustainable Future for World Heritage*, World Heritage Papers No. 13, Paris: UNESCO pp. 43–44.

Nicholas, L.N., Thapa, B. and Ko, Y.J. (2009) 'Residents' Perspectives of a World Heritage Site: The Pitons Management Area, St Lucia', *Annals of Tourism Research*, 36 (3): 390–412.

Robinson, M. (2000) 'Collaboration and Cultural Consent: Re-focusing Sustainable Tourism', *Journal of Sustainable Tourism*, 7 (3): 379–396.

Robinson, M. (2011) 'Meaning in Chaos? Experiencing Cultural Heritage, Tourism and the Challenge of the Popular'. In Halbertsma, M., Stipriaan, A. van and van Ulzen, P. (eds) *The Heritage Theatre: Globalisation and Cultural Heritage*, Newcastle: Cambridge Scholars Press, pp. 215–239.

Robinson, M. and Silverman, H. (2015) 'Mass, Modern and Mine: Heritage and Popular Culture'. In Robinson, M. and Silverman, H. (eds) *Encounters with Popular Pasts: Cultural Heritage and Popular Culture*, New York: Springer, pp. 1–30.

Rössler, M. (2006) 'World Heritage Cultural Landscapes: A UNESCO Flagship Programme 1992–2006', *Landscape Research*, 31 (4): 333–353.

Rössler, M. (2012) 'Partners in Site Management. A Shift in Focus: Heritage and Community Involvement'. In Albert, M.T., Richon, M., Viñals, M.J. and Witcomb, A. (eds) (2012) *Community Development through World Heritage*, World Heritage Papers No. 31, Paris: UNESCO, pp. 27–31.

Salazar, N. (2005) 'Tourism and Glocalization: Local Tour Guiding', *Annals of Tourism Research*, 32 (3): 628–646.

Su, M.M. and Wall, G. (2012) 'Community Participation in Tourism, Mutianyu Great Wall, Beijing, China', *International Journal of Tourism Research*, 16 (2): 146–156.

Sullivan, S. (2004) 'Local Involvement and Traditional Practices in the World Heritage System'. In de Merode, E., Smeets, R. and Westrik, C. (eds) *Linking Universal and Local Values: Managing a Sustainable Future for World Heritage*, World Heritage Papers No. 13, Paris: UNESCO, pp 49–55.

Taylor, K. and Lennon, J. (2010) 'Cultural Landscapes: A Bridge between Culture and Nature?', *International Journal of Heritage Studies*, 17 (6): 537–554.

Von Droste, B. (2012) 'World Heritage and Globalization: UNESCO's Contribution to the Development of Global Ethics'. In Albert, M.T., Richon, M., Viñals, M.J. and Witcomb, A. (eds) *Community Development through World Heritage*, World Heritage Papers No. 31, Paris: UNESCO, pp. 11–15.

Wall, G. and Black, H. (2004) 'Global Heritage and Local Problems: Some Examples from Indonesia', *Current Issues in Tourism*, 7 (4/5): 436–439.

Waterton, E. and Smith, L. (2010) 'The Recognition and Misrecognition of Community Heritage', *International Journal of Heritage Studies*, 16 (1/2): 4–15.

World Heritage Committee (2002) Budapest Declaration on World Heritage, CONF 202 9, Paris: UNESCO. Online, http://whc.unesco.org/en/decisions/1217/2002 (accessed December 2014).

World Heritage Committee (2007) Thirty-First Session, Christchurch, New Zealand, 23 June–2 July WHC-07/31.COM/24, Paris: UNESCO. Online, http://whc.unesco.org/archive/2007/whc07-31com-24e.pdf (accessed December 2014).

2 World Heritage as a revitalization movement

Managing local and global tourism in UNESCO's heritage-scape

Michael A. Di Giovine

More than 40 years after the World Heritage Convention was signed, UNESCO's (United Nations Educational, Scientific and Cultural Organization) program designating places of local interest into sites of universal value has had an immense effect on preservation, tourism, and development initiatives. What was initially a network of 12 primarily Western sites has grown to become a worldwide social system, a *heritage-scape* (Di Giovine 2009a) composed of more than 1,000 places in 163 nation-states representing diverse cultures, times, and spaces across the globe. Although in its early years UNESCO operated under a decidedly Western notion of heritage that coincided with, and dialectically affected, the birth of nation-states and nationalist ideologies (cf. Garcia-Fuentes 2010; Ashworth 2008; Edwards 2007: 27) – a notion of heritage, UNESCO admits, that has been largely 'confined to architectural monuments', which 'over-represented' European and Christian structures (1994: 2–3) – today islands, marine areas, historic industrial edifices, and cultural sites in developing countries represent the largest increases in designations. Private trusts, national funds-in-trust and international consortia of experts operate in every continent. UNESCO's own World Heritage Fund, which assists locals in developing management plans, and provides 'technical assistance' and best-practices exchange programs for site managers in developing areas, has increased from disbursing \$2,500 in 1978 to over \$4 million in 2009 (World Heritage Centre 2010). And despite wars, terrorism, and cyclical economic downturns, tourism continues to be the largest and fastest-growing industry in the world, surpassing even that of the oil trade (UNWTO 2007). Hundreds of millions of tourists visit World Heritage Sites annually (cf. UNESCO 2010); the private, user-generated website www.worldheritagesite.org even posts a long list of properties whose visitation rates exceed one million people per year. And while 2009 was the first year in at least two decades in which visitor arrivals did not increase, tourism to developing countries – particularly in Africa – continued to grow at a rate of 5 percent (Kester 2010: 5–6).

UNESCO and tourism development

The conjunction of historic preservation, tourism, and development had been present from the very beginning of UNESCO's initiative; indeed, one could say that it was born from it in the mid-1950s, when Egyptian president Jamal Nasser, eager to industrialize his country and to promote himself as the leader of Arab nationalism, set out to construct the Aswan High Dam – an act that would completely submerge the famed Nubian temples of Abu Simbel, cut into solid rock on the order of pharaoh Ramses II in 1244 BC. In 1960, UNESCO launched an unprecedented international campaign to move the temples to higher ground – a daring project that involved cutting the temples into 20-ton blocks and relocating them atop a man-made mountain overlooking the original site. Over 40 member-states and numerous private individuals raised more than the required amount of $42 million dollars; the effort also mobilized archaeologists from around the world, who worked side-by-side for six years in the Sudan's 'practically virgin land' (UNESCO 1980: 30). Abu Simbel had a profound effect on the future of World Heritage, and UNESCO proclaimed it 'a triumph of international solidarity' (UNESCO 1980). Through diplomacy in securing aid from member-states and awareness-raising publicity appealing to private individuals and non-profit organizations, UNESCO literally moved mountains. It revealed the strong emotional relationship that such heritage monuments – and their impending transience – exert upon people, irrespective of national origins. Today, despite their isolated location, Abu Simbel is one of Egypt's main tourist attractions; visitors trek to the top of the mountain in buses, caravans, and even private aircraft. Lastly, it showed how development initiatives could spur preservation and tourism, and, conversely, how today tourism and preservation have helped develop a once-impoverished area.

'Development' was always at the forefront of UNESCO's work. Founded in 1945, UNESCO was very much a part of the diplomatic and developmental initiatives of the immediate post-World War II era. This was a period marked by the Marshall Plan and Truman's famed Four Points, which were concerned with global reconstruction and development for international security interests. Truman stated that '*only by helping the least fortunate of its members to help themselves can the human family achieve the decent, satisfying life that is the right of all people*' (Truman 1949). While Truman's ideology emphasized a developed/undeveloped binary that located Western and non-Western nation-states along an unequal continuum of civilizational evolution (cf. Rist 1997: 72–73), four years earlier UNESCO had offered an alternative – one that had the capacity to elevate all cultures to the same valorized status. UNESCO's plan for creating worldwide peace rested on the creation of global selves in the myriad minds of the world's individuals. Noting that throughout history wars have always originated in some part through local suspicions or ignorance of other peoples at the individual level, UNESCO's Constitution posits that people's identities are problematically based on traditional territorial conceptions (UNESCO 1972: 1) that are constructed and diffused through these emotionally charged monuments: 'Since wars begin in the minds of men, it is in the minds of men that the defenses of peace must be constructed' (UNESCO 1945: 1). UNESCO's peacemaking endeavor thus involves a fundamental reworking of the international geopolitical system not through physical conquest but by reordering *individuals'* sense of place the world over – so that no longer do they base their identities on collective antagonism toward difference, but on the celebration and internalization of diversity at the individual level.

This goal manifested itself in the 'tourism development' initiatives of the 1970s – the same decade that the World Heritage Convention was passed – which were embraced by post-colonial states after the failure of Point Four-inspired industrialization plans. Without realizing the psycho-social tensions it would cause, these states recognized their competitive advantage to lie in the very underdevelopment they sought to transcend; tourism celebrates the very traditions that must necessarily be changed through modernization interventions. Tourism was also attractive because it was relatively inexpensive to utilize as a resource, it could bring in short-term cash flows, it could impel conservation and infrastructural development (UNESCO 1972), and it could produce both direct and indirect employment. But despite ambitiously altruistic aims, the *deus ex machina* conception of tourism and development was quickly found to be a problematic one (cf. UNESCO 1976). First there is the paradoxical role played by outsiders in both creating and helping to alleviate 'underdevelopment' in these areas; tourism development is also often conceived as a means of empowering and 'bettering' hosts that necessarily rests in the hands of the guests (Di Giovine 2009c: 211; 2010: 273). This outside-in, top-down structure frequently excludes locals in touristic and managerial interactions, and problematically undermines the very premise of UNESCO's program. *Betterment* does not simply mean economic improvement and increased employment, although that is often the case; it also includes 'empowerment' (Kane 1993; Rist 1997: 130), global 'integration' (Truman 1949; UNWTO 1974) and international valorization (Dure 1974; UNESCO 2005: 9), which can emerge precisely by adopting such a globally networked industry as tourism. Furthermore, this paradigm often treats culture as a resource to be economically exploited, rather than as the toolkit that society uses to organically make meaning of the surrounding world (Di Giovine 2010: 274).

The revitalization paradigm

Elsewhere I appropriated Anthony Wallace's classic 1956 model of revitalization movements to offer an alternative to the outside-in, top-down, and inorganic nature of the 'development paradigm' (Di Giovine 2009c, 2010). Revitalization, as Wallace describes it, is a model for understanding culture change, brought about by significant and often complex socio-biological shifts in the host society, which have been caused by new or intensified contact with an outside society. The society perceives itself 'organismically' – as a living ensemble of environmental, biological, social, cultural, and psychological factors, which has begun to sicken or otherwise veered off track. Visionary local leaders, wishing to restore the society's ideal cultural values (Wallace 1956: 265) will make recourse to the past – their cultural heritage – to point out what they view as the proper path to future well-being. While believed to be restoring their society to its former healthy state, this act creates a new system, which integrates into their ideological scheme new knowledge, new symbols, and new cosmological understandings that often stem from interactions with a diversity of outside forces.

We can already see how 'World Heritage' fits into this model: Espousing a narrative of 'unity in diversity' (Di Giovine 2009a: 119–144), World Heritage values culture 'in all its forms' (Matsura 2001: 11; 2002: 3). Former UNESCO Director-General Koïchiro Matsura writes: 'Culture is diversity, an infinite tapestry of distinctions, nuance and change; a relentless return to all that exists in order to render it both new and the same, to understand it and bring it to life. Diversity is essentially cultural, just as culture is diversity' (UNESCO 2002: 4). Invoking a cyclical notion of rediscovering or restoring the past, heritage discourses privilege past traditions, past creations, and past memories – and argue for its preservation from destructive outside forces that come with the passing of time (Di Giovine 2009a: 303–305; cf. Lefebvre 1991: 221; UNESCO 2002: 13). UNESCO's *World Heritage Information Kit* merges the dual concepts of monumentality as culture-preserving and the need to stave off time's destructive force of transience, when it asserts: 'Heritage is our legacy from the past, what we live with today, and what we pass on to future generations. Our cultural and natural heritage are both irreplaceable sources of life and inspiration. They are our touchstones, our points of reference, our identity' (2005: 3). But preservation is itself a creative act; it makes something new in meaning, if not form, in its effort to 'restore' an object back to an imagined past state. The same is true in a revitalization movement. Such restoration of cultural values is really an invention of a new route forward – a 're-invention of tradition' (Di Giovine 2010: 275; cf. Hobsbawm and Ranger 2012). 'Development' is not conceived of as a linear and evolutionary progression on the world stage, but as a cycle: one's culture can be germinated, born, flourish, grow sick, and become born anew (Wallace 1956: 267).

The theory behind UNESCO's World Heritage program also lends itself to a revitalization paradigm, for it privileges the individual over and above the collective, whose identities are often predicated on opposition to other groups (Di Giovine 2009a: 78). Its meta-narrative of 'unity in diversity' implicitly values not only the multiple meaning-making processes at work at any site, but also implies that such a conjunction ultimately creates something greater than the sum of its parts. A revitalization paradigm does not eschew the intervention of outside forces – which often come in the common form of a development 'plan' (Di Giovine 2009c: 218–221; cf. Norohona 1979: 9–10; Burns 1999), the hallmark of modern development theory (Ferguson 1994: 37; cf. Chambers 1994) – but views cultural regeneration as the totalization of multiple procedures that impels locals to make significant changes to their society by calling upon, yet reinventing, their narratively understood identity. Unlike a 'development' paradigm, the revitalization model brings local 'hosts' into a direct and more equitable relationship with touristic 'guests' as well as with the variety of other actors operating within what I have elsewhere called a 'field of touristic production' – a multi-layered complex of different subjective communities engaged in Bourdieuian struggles of positioning and position-taking to create, define, and promote formative encounters with place (Di Giovine 2009a: 42–48; cf. Bourdieu 1993). Through this process of mutual and multifarious interaction, contestation, and negotiation among a diversity of individuals and communities, locals drive their socio-cultural transformation by subtly expanding their very cosmologies and how they perceive their position in it.

The phenomenology of tourism is also extremely conducive to fostering a sense of 'revitalization.' As Graburn (1977) argued, tourism is a 'sacred journey' undertaken to experience a formative change from the everyday. Structured as a ritual (cf. Turner 1967), it serves not only as a *rite of passage* (van Gennep 1960) but also as a *rite of intensification* – 'periodic or cyclical rites that renew the social or natural world' (Graburn 1983: 12; cf. Chapple and Coon 1942: 485). Just as revitalization movements are responses to collective crises from their surrounding environment, so too are rites of intensification responses to 'group crises largely of an environmental character' (Chapple and Coon 1942: 507). It is no wonder, therefore, that tourism is, essentially, environment-oriented. No matter how active an itinerary may be, it is always conceived of as a transitory escape from the everyday, a time of rest, a time to recharge oneself and refocus one's eyes. While most theorists argue that mobility or geographic change is an absolute requirement of tourism – see, for example, UNESCO's first definition of a 'tourist' in 1963 (quoted in Norohona 1979: 3) – I have elsewhere argued that this change-in-place is essentially perspectival (Di Giovine 2009a: 145–149); it is less an embodied transformation of one's outside environment than it is an interior one. While he assumed global mobility in his theory of the 'tourist gaze', Urry (2002) nevertheless implied tourism's fundamentally perspectival nature: the tourist gaze is a socially organized process of seeing a place, which is 'constructed in relationship to its opposite, to non-tourist forms of social experience and consciousness' (2002: 1). This component is important, for it allows us to focus on the interaction itself, rather than on movement. Closing the 'host'/'guest' divide in traditional tourism theory (cf. Smith 1977), it also opens up the possibility that anyone – including locals – can seek the transformational intensification of tourism: all one requires is a shift of the gaze, a change in the way one approaches the site.

And on the practical level, while the heritage-scape can be considered meta-social, it is not constructed through a top-down approach. That is, with rare exceptions, a site must be nominated by the country in which it is found (UNESCO 1972: Articles 3, 4, and 5). That the nomination is in the hands of the State Party, rather than outsiders, is an important facet of UNESCO's non-conflictual and democratizing procedures. It is at once an attempt to engage and co-opt State Parties, as well as a powerful symbolic gesture by the nation-state of international solidarity. In my research in both Europe and Southeast Asia, more World Heritage Sites have been nominated through the endeavors of visionary local leaders than by the highest echelons of the nation-state; they often wish to obtain recognition and economic assistance from the nation-state, precisely by leveraging potential international valorization (Di Giovine 2009a: 216). In Việt Nam, for example, the imperial capital of Huế, the sanctuary of the minority Cham people of Mỹ Sơn, and the 'ancient town' of Hội An have been designated through years of politicking between local leaders (who solicited help from outside experts) and the national Communist Party.

Revitalization in central Việt Nam

To the government, who had just completed two wars against foreign 'imperialists,' the former imperial capital of Huế represented a distasteful mix of 'feudalism' – a Marxian mode of production ensurfing small peasantry (Tucker 1978: 53) – and collusion with Western imperialists, which was indicative of weak national leadership. (After seeing him at the World's Fair in Paris, Hồ Chí Minh had famously called the puppet emperor Khải Định a 'bamboo dragon'). Indeed, though Americans were responsible for bombarding Mỹ Sơn after learning the Việt Cong were using it to hide weapons, the Việt Cong were equally to blame for the destruction of Huế; during the Tet Offensive they went door-to-door in a rampage of destruction, and used the fortress for cover. After the war, the impoverished locals were allowed to use the remains for firewood and makeshift shelters.

In 1980, under the patronage of the Huế Relics Conservation Center – founded during the American occupation by high-status locals such as Pham Thuận An, a descendant of Khải Định's mother – a French archaeologist submitted a report to UNESCO detailing the damage to the city, which UNESCO then forwarded to the national government. Although the government had no use for Huế's narrative, and was engaged in an even more taxing war against the Chinese-backed Khmer Rouge, they sent architect Hoàng Đạo Kính with Nguyễn Xuân Đồng – an archaeologist who was once the curator of the Đà Nẵng Museum, which houses artifacts from Mỹ Sơn and other local Cham sites – to Poland and Hungary for technical assistance. In their travels, they met Kazimierz Kwiatkowski – or 'Kazim,' as the Vietnamese will come to affectionately call him. A conservationist at the Pracownie Konserwacji Zabytkow (PKZ) – known in Việt Nam as the Ateliers for the Conservation of Cultural Properties of Poland – Kazim was persuaded to move to Huế to assist the local conservation committee compile an 'action plan' for conservation (finalized in 1981), and to help preserve the badly damaged temples of Mỹ Sơn.

Sometime around 1983, while Kazim was working in Mỹ Sơn, he heard from some Soviet tourists of a nearby town so perfectly preserved in its traditions that it resembled an 'old Hà Nội'. This was Hội An. According to one local government official, who was present at the time, 'he was so surprised to see the old buildings' that he came back every weekend and urged the Ministry of Culture to conduct their own management studies (Di Giovine 2009a: 263). Kazim's visit roughly coincided with a research trip by Daniel Robinson, a writer for *Lonely Planet*, who would compile the first edition of a guidebook on former French Indochina; Robinson would end up dedicating a full fourteen pages to this small and relatively unknown town – stunningly, the same amount of space that he gave to Saigon and Hà Nội. When the book was published in 1991, backpackers arrived in such numbers in the small town that the Ministry of Culture and Information almost immediately established a program of tourist services there.

Locals such as the carpenter Tấn Ký realized the value in opening their homes to these visitors; charging a small fee, Tấn Ký was able to raise funds to restore his two-level wooden home. Tourism bestowed not only economic value, but also psycho-social value, to his family and his heritage; today his children proudly exhibit the family's heirlooms, period furniture, and black-and-white photographs in the home. Others found it more difficult to restore their dilapidated houses, and preferred to tear them down and erect new ones in concrete and tin. Visionary local leaders once again acted as mediators between locals and the nation-state; they secured help from preservationists to restore a few exemplary edifices at no cost, they distributed licenses for locals to work in the tourism sector, and they provided limited grants for local restoration and tourism integration projects. These local leaders, who espoused indigenous worldviews, also mediated between insiders and outsiders, and worked in close partnership with preservationists and advisory bodies, to nominate the site as World Heritage. Importantly, they called on outsiders to assist, rather than initiate, the development plan – producing, in the end, a new system of values, structures, and aesthetics that makes recourse to the society's unique past.

'Transitive revitalization' in southern Italy

An ethos of revitalization, cultivated through tourism and preservation, can also germinate in one place and inspire World Heritage initiatives in another. One such example is Benevento, Italy – once the capital of the pre-Roman Samnite people, a prosperous Roman center of trade on the Appian Way, and the longest-existing duchy of the Lombards in Italy. The only land-locked provincial capital in the Campania region, Benevento is historically one of the region's poorest (Davis 1998) and resistant to change, notwithstanding intense government intervention since the *Risorgimento* (Schneider 1998). Despite outstanding monuments to its past history – including an exceptionally well-preserved Triumphal Arch of Emperor Trajan, thirteenth-century Romanesque bronze doors, and Santa Sofia, one of the few Lombard circular churches in existence – few tourists travel to the city. Eight kilometers away, however, is the small and formerly impoverished village of Pietrelcina, the birthplace of the immensely popular twentieth-century Capuchin stigmatic and saint Padre Pio of Pietrelcina. Although Pietrelcina welcomes only 10 percent of the roughly six million pilgrims who visit Pio's tomb and shrine 165 kilometers away in San Giovanni Rotondo (*Gazzetta del Mezzogiorno* 2009), such visitation valorizes Pietrelcina as a site worthy of seeing, and, locals believe, a site specially blessed by Pio (Di Giovine 2009b: 483; 2012c: 119–123). It has fueled a revitalization movement that has not only materially impacted the aesthetics of the town, but also served to regenerate its very identity (Di Giovine 2009d). Reacting to a mix of extra-local pressures and the prospect of material 'betterment' from tourism, the town has reformulated its idea of heritage in connection with Pio's biography – which includes valorizing (imagined) aspects of everyday life at the time of his youth, from the food produced to the complete restoration of the city center (Di Giovine 2010; 2012a; 2014a; 2014b). Yet revitalization is not simply resting on the past, stagnating it; rather, it is about calling upon locals' worldview and values to improve society in relation to outsiders. For nearly a decade, Pietrelcina's town council and its *Pro Loco* (tourism promotion organization) have secured funding from the region, nation, and even the European Union to conduct excavations, restore the town center, open sustainable agritourisms, and host numerous international festivals nearly every night of the summer. These are clearly for the benefit of locals, though they reach beyond traditional sensibilities; I asked the mayor if the locals even liked jazz – the subject of one such public concerts – and he candidly replied 'probably not, but it's important to bring culture to them' (Di Giovine 2012b: 529).

This revitalization has rippled throughout the region, inspiring a renewed local interest in Benevento's cultural attractions. Archaeological excavations and historic preservation initiatives are sweeping through the city for the first time since post-World War II reconstruction; streets are being paved and pedestrianized; and, more importantly, many citizens have begun to declare their hopes that the city will be added to Italy's long list of UNESCO sites. Local historians and archaeologists have been given a platform for extolling the universal significance of Benevento's monuments, and secured regional and national support to nominate Benevento's church of Santa Sophia – one of only a few Lombard churches with a circular floor plan – within UNESCO's 2011 multi-city World Heritage designation 'Longobards in Italy: Places of the Power.' This designation integrates the city into a valorizing narrative that sees it as 'one of the main roots of the beginnings of the medieval European world and the establishment of Western Christianity' (UNESCO 2011: 221; cf. 'Italia Langobardorum' 2010; Santillo 2004).

Conclusion

Like that of Việt Nam, the case of Benevento reveals that UNESCO's meta-narrative inspires individuals confronting culture change and outside pressures on their ways of life; locals embrace the heritage-scape's potential not merely for economic benefit, but also for valorizing their history and culture as worthy to be counted on the world stage. It also considers tourism's phenomenological aspects as a formative and revivifying ritual that bonds locals and outsiders in a special *communitas* and is based upon deeply held narrative claims. It is thus inadequate to view the diverse touristic initiatives through a developmental paradigm for the same reason: although clearly outside forces – from national and international grants to revenue from international visitors – have helped fuel these new projects, local practitioners appropriate the initiatives to drive the movement rather than merely 'participate' in it (Di Giovine 2009c).

All of these cases represent how a UNESCO designation, coupled with the revivifying and perspectival interaction we call tourism, can fuel a revitalization movement that creates a more inclusive and sustainable form of socio-cultural 'betterment' than a traditional tourism development paradigm. It allows for greater attention to be paid to the significance of locals' discourses and practices in relationship to the panoply of other actors, and is sensitive to alternative value systems while also being responsive to the real material impacts of global forces on the local. Focusing on the transformations brought about from within a culture, in reaction to and in conjunction with outside forces, the revitalization paradigm can better capture the complex processes of negotiation that occur within a touristic field of production, as well as the variety of intertwining elements that inevitably factor into any 'culture change' phenomenon. Most importantly, it provides a way to move beyond the tautology of development theory, opening the way for alternative views of how individuals and groups work to 'improve' themselves as active agents, and how outsiders can productively contribute to the process on a more equitable basis. It is in this way that sustainable 'peace in the minds of men' can truly be created.

References

Ashworth, G.J. (2008) 'Heritage: Definitions, Delusions and Dissonances.' In R. Almoêda et al. (eds.) *Heritage 2008: World Heritage and Sustainable Development*, Lisbon: Green Lines Institute for Sustainable Development, pp. 3–9.

Bourdieu, P. (1993) *The Field of Cultural Production*, New York: Colombia University Press.

Burns, P. (1999) 'Paradoxes in Planning Tourism: Elitism or Brutalism.' *Annals of Tourism Research*, (26) 2: 329–348.

Chambers, R. (1994) 'Participatory Rural Appraisal (PRA): Analysis of Experience.' *World Development*, (22) 9: 1253–1268.

Chapple, E.D. and Coon, C.S. (1942) *Principles of Anthropology*, New York: Holt.

Davis, J. (1998) 'Casting off 'The Southern Problem': Or, the Peculiarities of the South Reconsidered.' In J. Schneider (ed.) *Italy's 'Southern Question': Orientalism in One Country*, New York: Berg, pp. 205–224.

Di Giovine, M.A. (2009a) *The Heritage-scape: UNESCO, World Heritage and Tourism*, Lanham, MD: Lexington Books.

—— (2009b) 'Re-presenting St. Padre Pio of Pietrelcina: Contested Ways of Seeing a Contemporary Saint.' *Critical Inquiry*, special edition 'The Curious Category of the Saint', (35) 3, Spring: 481–492.

—— (2009c) 'Revitalization and Counter-revitalization: Tourism, Heritage and the Lantern Festival as Catalysts for Regeneration in Hội An, Viet Nam.' *Journal of Policy Research in Tourism, Leisure and Events*, (1) 3: 208–230.

—— (2009d) 'Saints, Citizens, Pilgrims: Religious Tourism and Revitalization in Pietrelcina, Italy.' In Anna Trono (ed.) *Tourism, Religion and Culture: Regional Development through Meaningful Tourism Experiences*, Galatina (Lecce), Italy: Mario Congedo Publisher, pp. 239–260.

—— (2010) 'Re-thinking Development: Religious Tourism to St. Padre Pio of Pietrelcina as Material and Cultural Revitalization.' *Tourism: An International, Interdisciplinary Journal*, 58 (3): 271–288.

—— (2012a) 'A Tale of Two Cities: Padre Pio and the Reimagining of Pietrelcina and San Giovanni Rotondo.' *Textus*: *English Studies in Italy*, 1 (1) : 155–167.

—— (2012b) 'Making Saints, (Re-)Making Lives: Pilgrimage and Revitalization in the Land of St. Padre Pio of Pietrelcina.' Unpublished dissertation, Chicago, IL: University of Chicago.

—— (2012c) 'Passionate Movements: Emotional and Social Dynamics of Padre Pio Pilgrims.' In David Picard and Mike Robinson (eds.) *Emotion in Motion: Tourism, Affect and Transformation*, Surrey: Ashgate, pp. 117–136.

—— (2014a) 'The Everyday as Extraordinary: Revitalization, Religion, and the Elevation of *Cucina Casareccia* to Heritage Cuisine in Pietrelcina, Italy.' In Ronda Brulotte and Michael A. Di Giovine (eds.) *Edible Identities: Exploring Food and Foodways as Cultural Heritage*, Surrey: Ashgate, pp. 77–92.

—— (2014b) 'The *Imaginaire* Dialectic and the Refashioning of Pietrelcina.' In Noel Salazar and Nelson H.H. Graburn (eds.) *Tourism Imaginaries: Through an Anthropological Lens*, Oxford: Berghahn Books, pp. 147–171.

Dore, R.P. (1974) *The Prestige Factor in International Affairs: Proceedings from the 23rd Stevenson Lecture*, London: London School of Economics.

Edwards, P. (2007) *Cambodge: The Cultivation of a Nation, 1860–1945*, Honolulu: University of Hawai'i Press.

Ferguson, J. (1994) *The Anti-politics Machine: Development, Depoliticization and Bureaucratic Power in Lesotho*, Minneapolis: University of Minnesota Press.

Garcia-Fuentes, J.M. (2010) 'Recycling Heritage: On a Spanish–Catalan Process.' Paper presented at Heritage 2010 – International Conference on Heritage and Sustainable Development, Évora, Portugal, June 23, 2010.

Gazzetta del Mezzogiorno (2009) 'Sei milioni di turisti all'anno a San Giovanni Rotondo.' August 28, 2009. Online, www.sangiovannirotondo.it> (accessed 3 July 2010).

Graburn, N. (1977) 'Tourism: The Sacred Journey.' In Valene Smith (ed.) *Hosts and Guests: The Anthropology of Tourism*. Philadelphia: University of Pennsylvania Press.

—— (1983) 'The Anthropology of Tourism.' *Annals of Tourism Research*, 10: 9–33.

Hobsbawm, E.J. and Ranger, T. (2012) *The Invention of Tradition*, Cambridge, UK: Canto Classics.

'Italia Langobardorum. I Longobardi in Italia. I Centri di Potere. (568–774 d.C.).' (2010) 'Candidacy' website. Online, www.italialangobardorum.it (accessed 3 July 2010).

Kane, J. (1993) 'With Spears from All Sides.' *New Yorker*, September, pp. 54–79.

Kester, J.G.C. (2010) '2009 International Tourism Results and Prospects for 2010.' Report given at the UNWTO News Conference, Madrid, Spain: UNWTO, 18 January 2010.

Lefebvre, H. (1991) *The Production of Space*, trans. Donald Nicholson-Smith, Cambridge: Blackwell Publishers.

Matsura, K. (2001) 'The Cultural Wealth of the World Is Diversity in Dialogue.' In *Universal Declaration of Cultural Diversity* official packet, Paris: UNESCO, p. 11.

—— (2002) 'The United Nations Year for Cultural Heritage.' 20 March. In UNESCO, *The World Heritage Newsletter*, No. 34, March–April 2002.

Norohona, R. (1979) *Social and Cultural Dimensions of Tourism*, Staff Working Paper No. 326, Washington, DC: World Bank.

Rist, G. (1997) *The History of Development: From Western Origins to Global Faith*, London: Zed Books.

Santillo, R. (2004) *The Via Sacra Langobardorum: From Benevento to Gargano*. Benevento: Auxiliatrix.

Schneider, J. (1998) 'Introduction: The Dynamics of Neo-orientalism in Italy (1848–1995). In J. Schneider (ed.) *Italy's 'Southern Question': Orientalism in One Country*, New York: Berg.

Smith, V. (ed.) (1977) *Hosts and Guests: The Anthropology of Tourism*, University of Pennsylvania Press: Philadelphia, pp. 21–36.

Truman, H.S. (1949) 'Inaugural Address.' Online, www.let.rug.nl/usa/P/ht33/speeches/truman.htm (accessed 6 June 2008).

Tucker, R. (ed.) (1978) *The Marx–Engels Reader*, New York: W.W. Norton and Company.

Turner, V. (1967) *The Forest of Symbols: Aspects of Ndembu Ritual*, Ithaca, NY: Cornell University Press.

UNESCO (United Nations Education Scientific and Cultural Organization) (1945) *Constitution of the UNESCO*, UNESCO 1st General Conference, London, November 16, 1945.

—— (1972) Convention Concerning the Protection of the World Cultural and Natural Heritage, Article 1, November 16, 1972, Paris: World Heritage Centre.

—— (1976) 'A UNESCO Report: The Effects of Tourism on Socio-cultural Values.' *Annals of Tourism Research*, (4) 2: 74–105.

—— (1980) 'Nubia: A Triumph of International Solidarity. Official Inauguration of the Temples of Philae and the Twentieth Anniversary of the International Campaign to Save the Monuments of Nubia.' Booklet, March 10, 1980, Paris: World Heritage Center.

—— (1994) 'Expert Meeting on the "Global Strategy" and Thematic Studies for a Representative World Heritage List.' Addendum to the 'Report of the World Heritage Committee Seventeenth Session.' Phuket, Thailand, December 12–17, 1994.

—— (2002) *Cultural Diversity: Common Heritage, Plural Identities*, Paris: World Heritage Center.

—— (2005) *World Heritage Information Kit*, Paris: World Heritage Center.

—— (2010) 'Sustainable Tourism Programme', Paris: World Heritage Center. Online, http://whc.UNESCO.org/en/sustainabletourism (accessed 3 July 2010).

—— (2011) 'Decision 35 Com 8B.33.' In *Decisions Adopted by the World Heritage Committee at Its 35th Session, 19–29 June 2011*, Paris: UNESCO, pp. 221–222.

UNWTO (United Nations World Tourism Organization) (1974) 'Establishment of the World Tourism Organization', *Annals of Tourism Research*, 2 (2): 83–88.

—— (2007) 'Another Record Year for Tourism.' Press release, 29 January, UNWTO, Madrid. Online, www.unwto.org/facts/menu.html (accessed 27 June 2007).

Urry, J. (2002) *The Tourist Gaze*, London: Sage.

van Gennep, A. (1960) *Rites of Passage*, Chicago, IL: University of Chicago Press.

Wallace, A.F.C. (1956) 'Revitalization Movements.' *American Anthropologist*, New Series, 58 (2) (April): 264–281.

World Heritage Centre (2010) 'International Assistance – Statistics.' Online, http://whc.UNESCO.org/en/intassistance/action=stats (accessed 28 May 2010).

3 Responsible tourism and poverty

The porters of the Inca Trail

James Rollefson, Carolina Espinoza Camus and Alexandra Arellano

Tourism development associated with Machu Picchu World Heritage Site has become a major economic driving force in the region of Cusco, long marked by extreme levels of poverty, particularly in the rural mountainous regions. Beyond access to revenue issues, poverty is marked by substantial limitations in access to basic services and infrastructure. In the pursuit of social inclusion and mobility, these limitations can be even more important than the basic financial situation of the population.

With the dominance of both international tourism and poverty in the Peruvian rural highlands, it becomes highly relevant to consider the extent to which the characteristics of responsible tourism are being reflected in the local tourist industry and how the interplay with local conditions impacts local development and poverty relief. This chapter focuses on a study of the porters of the Inca Trail to Machu Picchu, who typically come from some of the poorest communities of the Andean highlands. The chapter builds on a recent article by Arellano (2011) that evaluated some of the general consequences of tourism for the porter communities through the lens of social exclusion. The chapter first puts the porters' working conditions and benefits in the overall context of poverty, social inclusion and well-being in Peru. An explanation follows of how these conditions have emerged and shows the major challenges facing the porters and their federation in trying to improve these conditions. The authors then provide a more general discussion on the porter families' access to improved basic services in the remote areas and show how they affect their well-being and socio-economic mobility. It is seen that only a sustained and multidimensional approach to relieving poverty in this region will have any chance of significant success. The chapter then identifies some of the absolutely critical issues that any comprehensive program to enhance the local population's well-being must include.

Measuring poverty in Peru

In efforts to provide a quantitative basis for the identification and intensity of poverty, a number of different frameworks have been developed (Alkire and Santos 2010). One of the most common approaches is the analysis of the monetary requirements for a person to be able to afford the food, clothing, shelter, etc.

needed to ensure basic well-being. According to Peru's National Institute of Statistics and Informatics (INEI 2012), a person is considered poor if he/she cannot afford this basic basket and is considered extremely poor if he/she cannot even afford that portion of the basket for food. For the current study, the INEI analysis is most useful in providing a perspective on the value of the porters' wages and the extent to which a porter's wages can meet his family's basic requirements.

In 2011 the basic basket's cost, i.e. the threshold below which poverty is considered to begin, was determined to be S/.272 per person per month at the national level of Peru. Much more relevant to this study are the figures for the rural highlands which were determined to be S/.200 monthly for the whole basket and S/.121 monthly for the food basket only. Below the capacity to disburse S/.121 per family member a household family is considered extremely poor. The INEI includes within its methodology an accounting of the monetary value of the products the family may produce and consume directly.

The severity of the poverty in the rural highlands in 2011 can be appreciated by INEI's statistics which show 23.6% of this population segment in Peru as extremely poor and 62.3% below the poverty line. For the people classified as poor, the average gap between the capacity to pay and the poverty threshold level is 21.3%; i.e. their capacity to pay for their essentials is on average (200) (1 – 0.213) = S/.157.4 per person per month. In concrete terms this affirms the assertion made earlier that the study framework must deal with poverty.

The most recent framework developed in Peru was inspired by the Multidimensional Poverty Index methodology used by the United Nations Development Program (INEI 2012). Vásquez Huamán (2012) adapted this framework for Peru such that existing INEI data from its annual National Household Survey (ENAHO) could be incorporated into a new qualitative multi-dimensional index of poverty for Peru. For comparative purposes on a macro scale this type of framework using data available from existing surveys is useful and affordable.

For an analysis limited to a region there appear to be substantial advantages to linking the qualitative variables or dimensions much more concretely to the causes of the poverty and to tailoring the variables to the specific regional issues. This flexibility is possible because in a regional rather than macro analysis there is no real need to define variables that are explicitly measured already through a national survey. The emphasis on cause and effect not only facilitates the analysis of data but also lends itself to determining the interventions or programs that can be more efficient and effective. Questions can be posed as to the critical factors affecting social exclusion and how it can be alleviated. To what extent do the citizens of a region feel that they can or cannot improve their lot within the socio-economic environment dictated to them by the availability of services and infrastructure?

The analysis with multiple dimensions lends itself to identifying combinations of initiatives that will be far more effective in enhancing socio-economic mobility than a massive program dedicated to one specific issue such as income. Access to basic services such as education, health and infrastructure enriches therefore the condition of emergence and resurgence of poverty.

Analysis of the porters' working conditions and benefits

Tourism in Peru has only developed as a major economic driving force in the last 30 years. It was spurred on by the designation of the archaeological complex of Machu Picchu as a United Nations Educational, Scientific and Cultural Organisation (UNESCO) World Heritage Site in 1983 and the feedback of the first tourists, who overwhelmingly were able to appreciate its aesthetic value. Its reputation grew even further with its designation as one of the seven new wonders of the world. Tourism became relevant to the poor subsistence farmers of the highlands when trekking proved to be one of the popular options. Most notable was the four-day trek through the mountains ending up with a passage through the Inca gateway to Machu Picchu. The ruggedness of the terrain, the climate and the altitude (one pass is 4200 meters above sea level) made it imperative for the tourist agencies and guides to employ porters to carry the camping supplies. The farmers in the villages surrounding Machu Picchu had the necessary stamina and high-altitude adaptation to work effectively in conditions where few others could. Unfortunately the long tradition of exploiting the highly vulnerable farmers continued, and the farmers found it very hard to defend themselves as there were virtually no other opportunities for remunerative employment. The exploitation included poor wages, horrible working conditions and very heavy loads to carry. This culminated in a major strike and social unrest as porters clamored for more equitable treatment. In late 2001 Peru's National Congress responded to the obvious destructive exploitation by passing a law explicitly setting out minimum wages and working conditions for the porters.

An ominous sign of how difficult it would be to have the 2001 Porters' Law respected is found in a letter directed to the management team of the Machu Picchu Sanctuary (UGM) by the Association of Tourist Agencies of Cusco (AATC): "as a consequence of the agreements formulated in a whole sequence of meetings called by the DRITINCI and the UGM our institution is recommending to its members the minimum pay of 100 new soles for the services of a porter for the four-day trek in the Inca Trail . . . It is evident that there are only a few isolated cases of agencies not meeting the minimum pay as set down with the porters." This letter is incredible. The minimum salary set down within the new Porters' Law was 148.80 new soles in 2002 and the Association simply refused to acknowledge it. It is not surprising that since then there has been a constant struggle of the porters' unions and its federation to have this law respected. From a historical perspective then, it could reasonably be concluded that the foundations for responsible tourism in the area are very weak.

The minimum wage for a porter for 2012 was set at 175.20 new soles for the four-day trek. A number of tourist agencies pay on or about this salary, but others pay as little as S/.120 to 130. The opportunity to work is seasonal. The tourists are less frequent in the rainy season from November through March. In February the Inca Trail is closed for maintenance. In the high season the porters with solid ties to the agencies may make four and exceptionally five treks monthly. In the low season this can drop to one or two treks. Many porters work far less often than that because there are about six porters seeking work for every one porter position open when evaluated on a permanent job basis. (In approximate terms there are some 200 porters per day who start the trail for four days' work over a season of some 300 days. This means there are 240,000 work days available and so, considering a permanent job to have some 240 work days, there are available jobs for 1000 porters on a permanent basis. However, in 2012 there were some 5800 porters.)

The struggle for full compliance with the Porters' Law continues as conditions for many porters remain poor. The porters are not carrying the 40–45 kg of the first years but they do still often carry 25 to 30 kg and many are poorly fed and mistreated. This varies greatly from tourist agency to agency.

A study in 2007 evaluating 673 porters in relation to 36 travel agencies found that only three of the businesses respected the Porters' Law – the great majority of the agencies did not provide the minimum levels of food, rest time and salary, at the same time as not providing adequate equipment for cold nights (Direccion Laboral de Inspeccion del Trabajo 2007). A number of the better travel agencies are indeed respecting the law in most regards (the exception is most often the weight load limit where they use a 25 kg limit rather than the 20 kg limit set down by the law). The companies that largely respect the law tend to draw their porters from villages where the men have a long tradition of working as porters. Such villages are Huilloc and Patacancha in the valley above Ollantaytambo, and other villages near Chinchero. Agencies that pay some 10 to 25% less than the minimum wage and impose heavier loads and poorer working conditions exploit the more vulnerable porters from remote locations.

If an average of about 2.5 to 3 treks per month for a porter is estimated, the monthly wages amount to some 500 new soles. With the threshold for poverty set at 200 soles per month per person, this means that the porter can escape poverty if he is married with no children. With one or more children he will have considerable trouble making ends meet. The majority of mature porters have from three to five children. With three or more children they are well into the category of extreme poverty (five people at 115 soles per person at the extreme poverty threshold is 575 new soles).

This analysis is borne out time after time in the interviews with porters. The vast majority say that the salaries that they earn allow them to cover only the very basic elements for survival. Only a few young bachelors claim to live reasonably well. Unfortunately the wages the porters receive are still substantially larger than those offered by other opportunities they have in the region, so the labour market pressure is actually to lower wages rather than increase them.

There are some possibilities for a porter to improve his lot within the system itself, i.e. some social mobility is open to him. If he learns to speak reasonable Spanish and he shows good organizational skills and is proven to be responsible, he can become a chief porter responsible for organizing the other porters. He will earn some 10 to 20% more for this. To get to be a cook on the trail with considerable more earnings, the porter can learn the trade as an assistant cook and complete his training in a formal cooking school. This economic hierarchy is also reflected in terms of prestige. The chief porter wants to be recognized as such and not as a simple porter.

The report card on responsible tourism

In the framework of the Cape Town Declaration on Responsible Tourism (2002) the tourism dedicated to local development would: generate greater economic benefits for local people and enhance the well-being of host communities; improve working conditions and access to the industry; involve local people in decisions that affect their lives and life chances; and be culturally sensitive, engender respect between tourists and hosts, and build local pride and confidence. A review can be made of how the tourist industry has embodied the principles of responsible tourism with respect to local development. However, the caveat must be applied that there are some good tourist agencies that can be far from the norm described below.

The economic benefits are real but they are often begrudgingly given to the extent that most porters' families remain in poverty or extreme poverty. A small minority of the agencies have invested in the local communities. In some areas the initiatives of non-governmental organizations, tourist agencies and new government programs are making a significant difference. The working conditions for the majority of porters can best be described as poor, despite being better than ten years ago. The vast majority of agency owners remain distant from the porters and hardly ever involve them in any decision-making process.

A determining factor in setting the work conditions and benefits has been the huge imbalance in power between the agencies and the porters. Basically there are far more people seeking work than there are positions open to them. The porters won a much improved legislative framework for setting working conditions after their strike in 2001 but their euphoria was short-lived, as noted earlier. The federation set up after the strike to defend the porters' rights has been ineffective, the main reason being the power imbalance between the federation and the agencies. The federation also would have needed much better logistical and financial support (how do you get significant funding from extremely poor people?) and its executive would have needed organizational skills they have simply not been trained to acquire. Furthermore, they have tried to run their operations with extremely limited professional support. Up until 2012 there was no accounting system in place to give some transparency to the federation's operations. This has resulted in a loss of credibility and of confidence in the federation by the porters themselves. Only in 2012 were serious efforts made to reverse this situation. In summary, the porters would have needed exemplary cohesion and an extremely well-organized federation to defend effectively their cause over the last 11 years.

Many agencies methodically exploit the vulnerability of the porters as they implicitly or explicitly indicate that if a porter doesn't want to accept the conditions offered, there are many other porters willing to accept them. Even today this argument is highly successful in keeping porters under the toe of some agencies. Under this coercion, many porters sign blank forms and receive salaries well below the minimum, and the agency fills in the document confirming that the porter has indeed been paid the salary according to the law.

Initiatives have to be undertaken to make it socially and culturally unacceptable to exploit porters in this way. The responsible travel agencies need to be publicly recognized, praised and presented as the new social norm for the region. The immediate focus of the federation executive must be on strengthening its organizational capacity, working more productively with government authorities in getting sanctions applied to agencies for breaking the law and diffusing well-backed information that makes far more transparent the differences between the good and bad practices of the agencies.

The porters and their social environment

An analysis of the porter interviews showed that the social environment has played a very significant role in determining the well-being of the porter and his family. With the current major transformation taking place in Peruvian society, there is a wide range of scenarios. In the worst scenarios the porters come from remote locations. Their educational opportunities have been limited by childhood malnutrition, by the long distance to school and by the school itself that would consist of a one-room building with a tutor teaching all grades. The porter's family would have had a small plot of land to grow one or two staple crops such as potatoes. There would be no health center nearby and family numbers above the national average. Under these circumstances it becomes more understandable why so many children are abandoned to one of the many institutes operated by benevolent organizations around Cusco. Frustrations and poor health conditions encourage alcohol consumption and domestic violence. Spouses earn a little extra money by making and selling handicrafts. Unfortunately the tourist market for these goods has become saturated in the past few years and profit margins are so slim that it is hardly worth the effort any more. Many of the older children abandon school early and can therefore hardly speak any Spanish. So the vicious cycle repeats itself, with the adolescents escaping home to become porters and earn what appear to be good wages.

In the best scenarios the porters have worked their way up the ladder to become cooks. They work for a good agency and are paid a salary a little above the minimum set by law. They have not felt put down because of their culture or early hesitancy with Spanish. They are encouraged by the agency to continue in various training programs to help prepare for retirement. Their working conditions and equipment are adequate and they have much fewer health problems. The improved social context helps many to stop overdrinking. A new health center in the village provides basic health services to the porter's family. A nurse provides a valuable extension service teaching preventive medicine, family planning and hygiene that is much appreciated. The children use toothbrushes regularly and at the periodic dental inspection at the local health center they are proud to show off teeth with far fewer cavities than others. Parents encourage the children to do well in school and participate in community efforts supported by a non-governmental organization to ensure that the children have a good nutritional base at school and at home.

With the help of a charity and a good tourist agency that believes in supporting the porters in their villages, a secondary school has been built to complement the elementary school with at least one class for every grade. Because the road to the village has been improved considerably in the last couple of years, the professors come regularly to their classes and do not skip a Friday or Monday as they often did in the past. Some professors even live in the village. The elementary school has now implemented the bilingual program in which the students begin their education in their native Quechua and then Spanish is brought in with second-language training a few years later. The village even has a little college so that the costs of pursuing a higher education have been significantly reduced.

The provincial government has listened to the local population and has improved the roads considerably and built a soccer pitch and a cemented floor for practicing volleyball and basketball. One teacher has stepped in to teach volleyball, another basketball. The children are making good use of the facilities and appear healthier.

The home environment has changed considerably; cement floors have been put in and the area of the houses expanded. They have water piped into the house now and they even have a sewer system. The water looks good but still contains parasites so they have learned to boil their drinking water regularly. They now have an improved wood stove that is about three times more efficient than the old kind, and all the smoke from the new stove is vented to the outside. Because of this the children have far fewer pulmonary problems than in the past. The children are washing up more often as the importance of hygiene has become clear to them. However, it still isn't ideal because it feels like just too much trouble to heat up the water often to have a good hot shower or bath. They usually make do with a warm facecloth or a quick splash in cold water to clean themselves. Still, they have virtually eliminated scabies and other skin problems from which they suffered in the past. The house becomes cold quickly after sundown (below zero in winter) so they finish off their homework early and escape from the cold under a few wool blankets.

With all these improvements, especially with the improved education, the children can aspire to a much better lot than in the past. Some still feel very sad because the work opportunities, other than as porters, are virtually nil so the children will eventually seek their fortune in the cities. Their ties to the culture will surely be weakened and the parents regret this.

It is a fair approximation to say that there are at least some porter families experiencing each and every one of the possible scenarios between the two extremes given above.

This is all in flux. The very latest assessment (Feb. 2013) of the economic transformation of the area is that some 1000 fewer people have applied to be porters (comparing 2013 to 2012) as new and better work opportunities appear as a result of the development initiatives of the public and private sector.

Conclusion

In the case of the porters it is abundantly clear that their poverty is directly related to both their monetary income and their families' access to good health and educational services. Because of the lack of alternative remunerative employment the porters are extremely vulnerable, and many of the agencies exploit this by offering poor working conditions and salaries. Families in some of the porters' villages have recently improved their lot considerably as a result of a combination of better work conditions and benefits offered by some tourist agencies along with improved socio-economic conditions in the villages supported by charitable organizations, a few tourist agencies and government programs in infrastructure development. It is the combination of these efforts that have spelt success for some. In other areas the poverty conditions are like a highly interconnected net holding back progress. If just a couple of meshes are broken, the problems persist. Success is most notable when programs are coordinated so that a substantial hole in the net is cut for them to break out of poverty in all its dimensions. For instance, effective educational training has a number of necessary components: improved nutrition, appropriate language training commencing with elementary schooling in the child's mother tongue, better schools, and new attitudes to schooling from the parents and responsible and well-trained teachers. Without addressing all the issues, the programs will have limited success. Access to health advice and services are critical to the relief of poverty. A specific example showed the huge impact that effective family planning could make. In this region a small family can escape poverty, but add just two or three children and it slips into severe poverty.

The ongoing research-action project with the porters has profited considerably from the type of analysis presented here in order to work effectively towards sustainable, long-term solutions to the challenges faced by the porters.

References

Alkire, S. and Santos, M.E. (2010) Human development research paper, 2010/11, Acute Multidimensional Poverty: A New Index for Developing Countries. Online, http://hdr.undp.org/en/reports/global/hdr2010/papers/HDRP_2010_11.pdf (accessed August 2012).

Arellano, A. (2011) 'Tourism in Poor Regions and Social Inclusion? The Porters of the Inca Trail to Machu Picchu', *World Leisure Journal*, 2 (53): 104–118.

Direccion Laboral de Inspección del Trabajo (2007) Informe anual de la inspección del trabajo en el Peru – 2007. Online, www.mintra.gob.pe/contenidos/archivos/inspecciones/informe_anual_inspecciones-2007.pdf (accessed June 2012).

INEI (Instituto Nacional de Estadisticas y Informatica) (2012) Estadisticas de pobreza. Online, www.inei.gob.pe/ (accessed June 2012).

Vásquez Huamán, Enrique (2012) *The Peru Poor Invisible to the State: Social Inclusion Pending July 2012*, Lima: Centro de Investigación de la Universidad del Pacífico.

4 Machu Picchu

An Andean Utopia for the twenty-first century?

Amy Cox Hall

> Because the patrimony is our identification, our richness, the patrimony is our antecedents, it is our pride, our identity, our insignia or I mean, it is everything, our patrimony there (Machu Picchu) it is fashioned, so I think that one should do everything, everything to protect it because it isn't going to die with us, but is going to pass from generation to generation, etc., etc., no?
>
> (personal interview, May 16, 2008)

Today Machu Picchu[1] is visited annually by more than 900,000 people.[2] In the city of Cusco it is a widely held belief that on the order of 95 percent of the local population lives off tourism. Much of this tourism may be attributed to the allure of Machu Picchu. The global business of tourism is considered by many to have an economic multiplier effect that ultimately benefits, directly or indirectly, the majority of Cusco's population. Curiously, it is also widely and simultaneously acknowledged that many groups are marginalized or adversely affected by the tourist industry (Flores Ochoa 1996). People both love and hate the inevitable paradox of Cusco as a historical site and tourist mecca. Unlike other tourist attractions that often have little meaning or interest for the local population, the archaeological site of Machu Picchu remains highly significant for the citizens of Cusco, acting as spatial, visual, tactile, and living proof not only of a national narrative of greatness but also of personhood.

In this chapter I focus on ethnographic work conducted with tour guides in Cusco, Peru, from 2006 to 2008,[3] to examine how Machu Picchu has come to be imagined as an Andean Utopia (see Flores Galindo 1986). For the tour guides, Machu Picchu represents not only a romantic and timeless notion of the nobility of Incan civilization but also the mythical home and oracle of the modern nation. The 'sanctuary' of Machu Picchu is not just a gold mine for Peru's tourist 'industry without chimneys;' it also represents a spiritual destination where global and national pilgrims are nourished and strengthened. Throughout this article I suggest that the tour guides learn about their Incan heritage and what it means to 'be Incan' through the business of tourism and their work as guides. I conclude that today's Andean Utopia is constructed in the quotidian practices of tourism and in response to, and within, neoliberalism and globalization.

In short, I argue that Machu Picchu embodies the Andean Utopia, but also the threat to that Utopia.

Temporal bridges, sentient entities, and self-fashioning at the sanctuary

In many ways the job of the tour guides to Machu Picchu is to nurture a temporal bridge to an imagined past, both for their clients and for themselves. The guides' role in conjuring a specific sensation and memory is critical for a successful visit to the ruins. Without the guides' temporal management, the ruins can become just rocks, old building foundations, and unused agricultural terraces. Through stories, the tour guides craft a romantic and New Age image of Machu Picchu as an Andean Utopia for the twenty-first century.[4] One guide stated that during her tours she tried to go back in time, using her imagination to achieve love and value for the Incan culture and to convey how much was sacrificed (interview May 21, 2008).

Tour guides to Machu Picchu first invite their guests to romantically envisage the site as it was five hundred years ago. The visitor is prompted to imagine the thatched roofs lashed onto the crumbling rocky edifices and to see the stepped terraces not as empty grassy knolls but as abundant harvests planted by their Incan ancestors. Visitors are asked to envision the royal estate of the Inca Pachacuti, with separate neighborhoods for the peasants and the nobility, the laborers and their rulers. Visitors are brought to a small garden and told to imagine indigenous specimens growing at the sanctuary. At the Intihuatana stone, visitors are urged to feel the stone's primal 'energies' and imagine sacred ritual acts performed by priests.

Part of the mystery and attraction of Machu Picchu is its imagined history. The visitor who is unable to cross this temporal bridge is often disappointed. Crowds often infringe on the visitor's ability to transport herself into a virtual past. The performance of spiritual and historical transcendence is lost on the visitor who cannot, or does not wish to, imagine the place in a romantic or New Age fashion, 'the way it was.' While there are other possible histories to be imagined for Machu Picchu, the tour guides I interviewed emphasized the romanticized past of the archaeological site as well as their own personal connection to the ruins as shared Incan heritage.

A sense of history and connection to an imagined past was critical for the tour guides' own fashioning of self and place. As one guide commented: 'Machu Picchu is identity, it allows us to remember who we are . . . it is that which we were, that fortifies us to remember our ancestors' (interview June 13, 2008). Another guide referred to his relationship with Machu Picchu as self-nourishment. 'I shouldn't feel bad about my humble origins but sometimes I feel guilty, isolated for being a guide because I have more. I try to have people [in Cusco] know their culture . . . Guiding is self-nourishment about this culture . . . it is the best therapy' (interview May 19, 2008). Understanding their Incan heritage, and belief in the power of its principles, were partially learned through their work as guides.

Many of the guides I interviewed had visited Machu Picchu in their youth. However, several of the guides confessed that they did not understand the place and were not moved by their first visit to the sanctuary (interviews May 16, 21, June 2, 6, 13, 2008). However, with time, education, and subsequent visits to the sanctuary, the guides admitted that they began to understand the place, to develop an intimacy with the site, and to feel a sense of mutuality and camaraderie with the mountain. Reina,[5] a guide who had worked for the INC and Peru Rail before guiding at Machu Picchu, felt that 'to be Cusqueño is to be one of the Incas, Gods of the sun . . . I now identify with the culture, before no, but now yes, because of knowledge' (interview May 21, 2008). Through her work in tourism, Machu Picchu came to life and, along with it, her sense of belonging and connection. Another guide, responding to the question 'why did you want to be a guide?', also emphasized the linkages between learning, knowledge, and heritage:

> Because suddenly I like history more, I like to identify myself with that which is mine, or it's like I know a little more, and suddenly I know more the tradition of the experiences of the ancient Cusqueños, so that I identified myself a lot and I very much liked to study history, archaeology, anthropology and from there to being a guide.
>
> (interview May 16, 2008)

All of the guides interviewed felt that a visit to Machu Picchu, whether for the first or the thousandth time, brought something new to learn and to experience. One guide likened it to a movie theatre, where the object of the gaze is always moving and always different (interview June 6, 2008). Another described Machu Picchu as a book: it ends, but re-reading it one discovers new things (interview May 26, 2008). Another described it as living identity (interview May 19, 2008). The guides personified the site, and talked of being 'in a relationship' with Machu Picchu and of constantly 'getting to know one another.'

A temporal bridge was forged through the tour guides' ability to narrate an imagined past. However, in the interviews with the guides, Machu Picchu was not simply an object upon which to project meanings or fashion personal identity. Instead, Machu Picchu was talked about as if it were a *persona*. Machu Picchu was not simply a geographic place, an example of a nation's past, or even an example of a region's heritage. Machu Picchu was conceived as a living entity and often described by the tour guides as a sentient being with distinct 'energies,' 'spirits,' 'vibrations,' and even 'emotions.'

> And how I continue telling you, Machu Picchu is something enigmatic, something beautiful, something that, for me, gives me a little fear at times being there, at times it impacts me so much that I forget things. Being at

> Machu Picchu at times I want to sit, relax, take a moment in peace being there. When we are with people, at times, that vibration is contagious that each one has of emotion . . . For me, Machu Picchu is something incredible. And enigmatic, like I told you.
>
> (interview May 15, 2008)

Throughout the interviews, Machu Picchu was consistently portrayed not as a 'thing' acted upon but as a 'thing' or being which acted upon them.[6] More than historical fact or fiction, the tour guides felt an animistic connection with the mountain.[7] Machu Picchu was not only a historical place or a part of an abstract national narrative of greatness; it was also a part of a relationship that fortified and nurtured the guides' sense of self. Guides said that they kept Machu Picchu in their hearts, and that a visit to the sanctuary was like a religious pilgrimage, or a visit to a friend, not a work destination. Both an imagined sense of history and an intimate relationship fashioned the Cusqueño tour guides' sense of themselves as descendants of the Inca.

Through the act of guiding, the tour guides learned about their heritage and history and thus learned to value what had been learned, becoming who they were born to be. Moreover, because of their connection with Machu Picchu, the guides felt that they had a clearer identity and sense of who they were and where they came from compared with other Cusqueños. Through the practice and experience of guiding, the tour guides developed an understanding of their Incan heritage and what it meant that they were 'a descendant of the Inca.' As one guide put it: 'Being Cusqueño is a distinction, it has its own history, culture, food; [it is] to feel part of a culture, a heritage. Being a guide is motivation to know more deeply for tourism' (interview May 19, 2008).

Tourism and the city of Cusco

In my interviews I asked the tour guides whether or not tourism to Cusco and Machu Picchu brought advantages or disadvantages. The response was not simple or unanimous. While there was no consensus about whether or not tourism was advantageous, all agreed that tourism had brought change. Cusco's 'culture' was changing and this was both lamented and applauded during the interviews. The perception of loss, particularly of long-held traditions, was often believed to be a result of increasing contact with outsiders, specifically through tourism.[8]

Many feared that locals were trying to imitate tourists, allowing their own customs to be shaped by outsiders. Not wearing the typical Quechua dress or drinking Coca-Cola instead of chicha morada[9] were the typical hallmarks called up in the interviews. Shifts in language use were also consistently cited during the interviews, with some guides arguing that Quechua use was strengthened by tourism and others suggesting it had been weakened.[10] Faulting the general lack of knowledge of history and self, the guides argued in various ways that it was easy to succumb to the temptations of outsiders. Drugs, poor behavior, bad manners, and prostitution were cited as examples of those with 'weak identities.'

> Tourism has changed Cusco . . . [The neighborhood of] San Blas is now not Andino. Instead of Andean food, people eat Swiss food. People stay here with their bad habits, take drugs, smoke, sell their shit. In the past the porters on the Inca trail were locals, now they wear modern clothes and are not from here. We don't have a strong identity and we are becoming cosmopolitan – we imitate habits of other places – we orient ourselves for the tourist . . . Man should live in harmony with nature, it is necessary to care for it and protect it. Should be revaluing it, it is sad that we are not protecting our patrimony nor our resources; it is painful to see this.
>
> (interview May 19, 2008)

Not every guide felt that tourism was a culprit in bringing bad habits. An equal number of guides believed that tourism had also brought change that was needed. Many guides attributed a greater tolerance for Peru's marginalized to tourism. Respect for Quechua speakers was believed to be a direct result of the respect and interest tourists had for Cusco's indigenous population. A heightened interest in the Inca ancestors and Cusco's history was also attributed to tourism. To some, tourism strengthened identity and was viewed as a vehicle to raise consciousness. Responding to the question of whether or not tourism had changed Cusco in terms of heritage, one guide felt that tourism had helped nurture the identity of Cusqueños.

> Because of tourism, not so much, before there was not as much identity. Now with tourism there is more identity, yes, because before those who spoke Quechua were marginalized, that which was 'pueblo' was marginalized . . . that has been since colonization there was so much marginalization, so much disdain, so much work as slaves for those they considered inferior, etc. That has been the case until very recently; recently national identity has grown and so that me, in my case for example, I feel proud to speak Quechua and so I go teaching and so I go telling my daughters, look, when I was a girl I traveled with my mom and I saw these things and now, look . . . me in my work I apply that, I tell them imagine all that I have been through, now I have to tell those things and for me it is part of my life, no? So I tell them look at these things; there is the truth! And so on, no? In that way then, they go on believing.
>
> (interview May 16, 2008)

Echoed throughout the interviews was a need for the Cusqueños to fortify themselves against the forces of change brought by tourism. The solution to a changing Cusco was a resounding and unanimous desire to improve tourism management and to 'fortify culture.' Through the strengthening of culture one could begin to respect Cusco as a community. The majority of tour guides felt that education was a primary way in which fortification could take place. Somewhat paradoxically, tourism and guiding represented one example of how to strengthen culture and identity through education. Guiding required and encouraged learning about Cusco's Incan past to craft a heritage that was consumable for tourists and guides alike.[11]

Conclusion

While some may critique the guides' dreams as capitalist fantasies, quaint folk iterations or New Age delusions, to take the guides' claims seriously means to reevaluate our notions of tourism and indigeneity and how the two are deeply intertwined. From the temporal bridges fashioned through Peru Rail's train journeys, to guiding at the sanctuary, the industry of tourism has participated in the construction of Machu Picchu as an Andean Utopia. In part through guiding visitors to their 'sanctuary,' the guides learned more about 'being Incan.' Machu Picchu has become part of the guides' sense of self and part of their dreams for the future. Paradoxically it is a place shaped and nurtured out of the complicit and tacit relationships sustained between Incanism, global tourism, New Age spirituality, and a neoliberal economic order. Framed as an antidote to an alienating commercial and global world, Machu Picchu has become the guides' twenty-first century Andean Utopia. This commercial and ritual pilgrimage to Peru's mythical home renews and fortifies the lives of guides. For the tour guides, their identity as 'being Incan' was partially earned and learned through tourism to Machu Picchu.

Although tourism management is acknowledged as a problematic tool for development, it was through tourism and difference that much was learned and revered about Cusco's reimagined Incan past. Both economic and moral-social development could occur through learning and understanding the region's Incan history. The references to history and collapsing of self into that history were part of the guides' work in Utopia dreams that involved the sanctuary. Machu Picchu was sacred because it represented physically and spiritually a path to truth. Machu Picchu was 32,573 hectares of pride and potential. It was a place with energies that, if you were attuned, could fortify your soul.

Notes

1 In 2009 the sanctuary of Machu Picchu was managed by four state entities – the National Institute of Culture (INC), the Regional Management of Foreign Commerce and Tourism (DIRCETUR), the National Institute of Natural Resources (INRENA), and the local municipality of Aguas Calientes.

2 According to the National Institute of Culture (INC), the total number of visitors (both national and international tourists) to Machu Picchu for 2007 was 439,122. In 1986, the number of visitors was 146,000 (Flores Ochoa 1996).

3 For this study I worked with the Colegio de Licenciados en Turismo (COLITUR). COLITUR is a professional association for those who have graduated with a bachelor's degree in tourism. Some estimate that there are currently 4000 guides actively working in Cusco and its environs. At the time of my research, COLITUR had nearly 1000 members and was rapidly growing. I formally interviewed 25 COLITUR guides and informally 25 more during waiting periods, visits to Machu Picchu, and bus trips. Each interview lasted between one and two hours. My only requirement to be interviewed was that the guides had experience working at Machu Picchu.
4 See Michael Hill (2007, 2008) for a discussion on mystical tourism in Cusco, Peru.
5 All names are changed.
6 Much has been written on sentient non-human being in the Andes (see Allen 2002; Abercrombie 1998; Platt 1997; Nash 1992; Taussig 1980).
7 See Marisol de la Cadena (2010) for her work on indigenous rights and mountains as sentient non-human beings.
8 However, some guides quickly pointed out that tourism was not the only culprit in cultural change. Several guides argued that while tourism needed to be better managed, many of the changes had to do with globalization in general, not tourism per se. 'It is not just tourism, but globalization that has changed Cusco' (interview June 6, 2008). The dozens of internet cafes, the constant blare of televisions, and technology in general were also included in what was fostering change in Cusco.
9 Chicha morada is a non-alcoholic refreshment made by boiling corn with cinnamon, oranges, and other spices.
10 Some of the guides pointed out that it was still not popular to learn Quechua. Instead, tour guides who did not speak Quechua focused on learning Japanese, English, or German. Still others felt that the resurgence of Quechua was due in part to the respect and admiration tourists had for the language (interview May 29, 2008).
11 See Barbara Kirshenblatt-Gimblett (2006) for a discussion on the connections between heritage and tourism.

References

Abercrombie, T. (1998) *Pathways of Memory and Power: Ethnography and History among an Andean People*, Madison: University of Wisconsin.

Allen, C. (2002) *The Hold Life Has: Coca and Cultural Identity in an Andean Community*, Washington D.C.: Smithsonian Institution Press.

De la Cadena, M. (2010) 'Taking Indigenous Politics in Its Own Terms Requires an Analysis beyond "Politics",' *Cultural Anthropology*, 25 (2): 334–370.

Flores Galindo, A. (1986) *Buscando un Inca: Identidad y Utopia en los Andes*, La Habana, Cuba: Casa de las Americas.

Flores Ochoa, J. (1996) 'Buscando los Espiritus del Ande: Turismo Mistico en el Qosqo,' in H. Tomoeda and L. Millones (eds.) *La Tradición Andina en Tiempos Modernos* Osaka, pp. 9–29.

Hill, M. (2007) 'Contesting Patrimony: Cusco's Mystical Tourist Industry and the Politics of Incanismo,' *Ethnos*, 72 (4): 433–460.

—— (2008) 'Inca of the Blood, Inca of the Soul: Embodiment, Emotion and Racialisation in the Peruvian Mystical Tourist Industry,' *Journal of the American Academy of Religion*, 76(2): 251–279.

Kirshenblatt-Gimblett, B. (1998) *Destination Culture: Tourism, Museums and Heritage*, Berkeley, CA: UC Press.

Nash, J. (1992) *We Eat the Mines and the Mines Eat Us: Dependency and Exploitation in Bolivian Tin Mines*, New York: Columbia University Press.
Platt, T. (1997) 'The Sound of Light: Emergent Communication through Andean Shamanic Dialogues,' in R. Howard-Malverde (ed.) *Creating Context in Andean Cultures*, Oxford: Oxford University Press, pp. 196–226.
Taussig, M. (1980) *The Devil and Commodity Fetishism.* Chapel Hill, NC: UNC Press.

5 Interrogating the 'universal' in St. Lucia's Pitons Management Area

Jennifer C. Lutton and Gregor Williams

St. Lucia's Pitons have long been contested space, stirring up debates over issues of national sovereignty; cultural, historic, and natural heritage; and development strategies. In the 1990s, conflict over the national significance of and rights to the Pitons generated protests centering on the development of the Jalousie Plantation Resort, which is situated between the Pitons. Since the inscription of the Pitons Management Area (PMA) as a World Heritage Site in 2004, these debates have reemerged with renewed vigor in response to intensified development of exclusive resorts in the Soufriere region (where the PMA is located) and throughout St. Lucia. Such responses commonly emerge in World Heritage contexts, which are 'most often a product of a much longer, socially and historically situated process of contested mediation' (Di Giovine 2009: 37). The PMA, like many other World Heritage Sites, constitutes a 'zone of contestation' (Appadurai and Breckenridge 1991: 407) in which debates animated by local and global actors over history, culture, and economic development flourish.

Intensification of mass tourism and the recent focus on heritage tourism are responses to neoliberal trade policies, which have eroded St. Lucia's (and other ACP countries') trading preferences with Europe, specifically for bananas. St. Lucia was the largest banana producer in the Windward Islands until the mid-1990s. Agriculture has since declined dramatically and has been eclipsed by tourism, which contributes 45.8 percent of St. Lucia's GDP (World Travel and Tourism Council 2011). In this political economic context, heritage articulates with tourism in ways that both accommodate and challenge tourism's economic dominance and its effects on social organization, culture, and identity. Heritage tourism is championed as an economic vehicle through which St. Lucians can project history and culture for tourists and St. Lucians on their own terms, but the more economically successful heritage tourism interests remain in private hands or, like the PMA, cater to a global tourist elite.

Linking heritage with tourism in small-island states like St. Lucia brings heritage into a direct relationship with neoliberal economic development directives of global financial agencies (e.g., the IMF and World Bank) which are reinforced by regional agencies (e.g., the Caribbean Development Bank) and local governments to spread foreign capital interests. This entails the transfer of heritage assets and the income they generate to a local and global elite. David Harvey argues that neoliberalism redistributes wealth and income to the 'upper classes or from vulnerable countries to richer countries' and hinges on 'accumulation by dispossession,' a process involving:

> conversion of various forms of property rights into exclusively private property rights; suppression of rights to the commons; commodification of labor power and the suppression of alternative forms of production and consumption; colonial, neo-colonial, and imperial processes of appropriation of assets; monetization of exchange and taxation, particularly of land.
>
> (Harvey 2007: 34–5)

We see this process unfold in St. Lucia's socioeconomic transformation from an agricultural to a service-based economy. Sugar was the island's main export crop during slavery through the post-emancipation period and until the 1960s, when bananas became the dominant export. Bananas were a staple provision that could be cultivated on small farms, unlike sugar which required large tracts of flat land, most of which was owned by St. Lucian elites or foreign companies. Importantly, the shift to profitable small-plot cultivation coincided with regional anti-colonial movements and became emblematic of national independence (St. Lucia became independent in 1979). St. Lucian author Anderson Reynolds describes the experience of economic and social achievement that accompanied the transition to banana farming:

> In as much as sugar was synonymous with slavery, bananas could be associated with freedom. Banana cultivation empowered the people because unlike laboring on the sugar plantations, farmers cultivating their few plots or acres of bananas had to become self reliant, responsible, forward-looking economic agents . . . Suddenly, farmers' grass and cow dung thatched huts gave way to wooden houses with galvanized roofs; mud floorings gave way to linoleum covered wooden floors; flambeaus and kerosene lamps to electric lighting; and donkey and horse drawn carts to automobiles. Suddenly, banks multiplied, primary schools and health centers appeared in the most isolated of villages . . . roads stretched over mountains, rivers and valleys, running water penetrated into the heart of the very hinterland.
>
> (Reynolds 2003: 94)

Many St. Lucians share Reynolds' observations about bananas' pivotal role in the development of the island's modern economy and national identity.[1] For those who benefited from the days of 'green gold,' the values generated by the Afro-Creole peasantry's history of resistance, its social structure and cultural practices, and its achievements are understood as under siege in tourism's political economic reconfiguration of society. Fear over losing the independence that many believe the banana economy provided evokes ambivalence towards tourism informed by a perception that the majority of tourism jobs restrict upward mobility and entail a step backward, when workers once again serve the whims of elite foreigners, aided by privileged nationals. The recent dedication of large sections of beachfront property – private and public – to tourism exacerbates this ambivalence and has renewed public protest in St. Lucia over the direction of economic development.

Given this historical trajectory, St. Lucians have been 'conditioned over long periods of time to respond to initiatives' imposed by external forces and entities (Friedman 1983: 152). Tourism expresses the structural, social, and economic continuities of inequality inherent in the plantation system, which created the conditions for St. Lucia's tourism industry, and which is dominated by outside interests, as well as by a small group of St. Lucian elites.

Who owns the Pitons? The consequences of 'universal' heritage

The PMA comprises an 11 square mile area in St. Lucia's southwest and includes seven zones with specified planning and management policies. The site incorporates private and public land, on which two volcanic plugs, the Pitons, are the most prominent feature, and encompasses the surrounding district of Soufriere, with a population of about 9,000.

A mulatto and white elite owned much of the Soufriere region during the colonial period, and some continue to control land now dedicated to tourism. Expatriates and international resort companies own resorts within the PMA that were former estates. Afro-Creole (and some Indo-Creole) peasant farmers from Soufriere and surrounding villages worked on these estates, as well as on their own adjacent small land holdings. Their descendants comprise the core labor force of the resorts that replaced the estates. Thus, race, class, and color hierarchies have historically shaped Soufrierians' tenuous and contestatory relationship to the territory within the PMA.

While there is broad consensus on the Pitons' significance to national heritage, the interpretation of this heritage as 'universal' was not without conflict. Efforts to obtain some variety of official heritage recognition for the Pitons have been ongoing since the mid-1980s, including two applications to the United Nations Educational, Scientific and Cultural Organization (UNESCO) for World Heritage inscription and plans to create a national park through government acquisition of estate land with assistance from the World Wildlife Fund. However, the United Workers Party government at the time agreed to a resort project that promised several hundred construction and hotel jobs. In 1992, an Iranian developer built the Jalousie Plantation Resort on three hundred acres of private land between the Pitons, destroying a good deal of unexcavated Amerindian and plantation period archaeological resources on the property. The resort's development brought competing ideas of heritage preservation and economic development into sharp relief in clashes between international environmental groups, St. Lucian NGOs, local intellectuals, the St. Lucian government, the developers, and ordinary St. Lucians. Former Labour Party prime minister Kenny Anthony (a law professor at the time) and St. Lucian Nobel Laureate Derek Walcott were among the protesters who argued that the Pitons and the land between them represented a contiguous area fundamental to national identity and sovereignty. In St. Lucian newspapers, Walcott called for the land's preservation for all St. Lucians, referring to the Pitons metaphorically as the *tete* (breasts) that feed the nation. This experience of heritage contestation, combined with concerns over the lack of access of many communities to tourism's economic benefits, led to the creation of the St. Lucia Heritage Programme in 1998.

In 2003, the PMA planning committee publicly launched their application to UNESCO on the Soufriere waterfront, which we attended. The Minister of Planning and Development from Soufriere challenged ideas of universality and declared at this event: 'We own the Pitons,' but suggested that World Heritage status would enhance tourism and proposed exploring ways to pay for maintaining the site through royalties from images of the Pitons used in advertising. The minister betrayed misgivings about the benefits of World Heritage status by reducing heritage preservation to a 'burden of cost' which the government must find ways to meet.

The Soufriere Action Theatre (SAT), a local drumming and theatre troupe, performed a satirical piece Derek Walcott wrote for a St. Lucian newspaper during the Jalousie controversy. Walcott's 'Litany to the Pitons' excoriates local political leaders and foreign capitalists who collude to dispossess St. Lucians of their heritage through economic development projects that jeopardize human and national development by offering low-paying, dead-end jobs to St. Lucians. Walcott writes:

May the Pitons pardon us

M Group [the development company] and Minister
More and more sinister
Manipulators
Merchants and traitors
Makers of waiters
Whores out of waitresses
Deep our distress is
Jalousie is one of
The seven deadly sins
Greed is another . . .
They would sell their own sons
They sold me, they sold you
When they sold the Pitons
May the next Generation
Curse a government so blind
It handed over a nation
Sealed, delivered and signed.

(Walcott 1992)

The performance made officials visibly uncomfortable, and SAT was proud to have upset them. A number of officials in attendance had protested against the resort and were now in positions of power where their ideological commitments, it seemed to SAT, took a backseat to their political ambitions. SAT contended that, compared to a national park, the PMA was an unacceptable compromise. The performance was a protest against erasure of controversial history where lands once privately owned by an expatriate and slated for public ownership became exclusive property for non-St. Lucians to enjoy and consume and on which St. Lucians labor.

Expansion and exclusion: resort tourism in the PMA

Intensification of tourism, integral to neoliberal plans that accompanied the end of trading preferences for bananas, has been particularly pronounced since the PMA's World Heritage designation and has been marked by an escalation of resort exclusivity in the Soufriere region.[2] The handful of resorts within and outside the PMA are small and high-end, with rates starting in the mid-\$300 range and going up to \$2,500 USD per night in high season. Following the PMA's acquisition of World Heritage status, two Soufriere resorts, Ladera Resort and Anse Chastanet (Jade Mountain), expanded. Currently, there are two major resort expansion plans. Anse Chastanet, which lies just outside the PMA, proposes constructing over 900 rooms in villas, condominiums, and hotel rooms. An expansion of Jalousie resort (to be renamed Sugar Beach) will develop 216 rooms in luxury villas and a hotel on 170 acres. The projects will add 1,100 more rooms to the 185 that currently exist between these two resorts, most likely producing lasting social, economic, and environmental effects in the Soufriere region.

It is important to note the character of the community in which this luxury resort development is occurring. In Soufriere and the surrounding districts of Choiseul and Canaries, from which the majority of the labor force come, most people live in small wooden frame or concrete houses, and a little less than a quarter use pit latrine toilets (St. Lucia Government Statistics Department 2001: 75). Many residents have only primary or some secondary school education and/or are low-skilled labor. Unemployment rates in the Soufriere region hover around 25 percent (St. Lucia Government Statistics Department 2004). There is, then, a stark contrast between the luxury resorts in which the very affluent tourists stay in the PMA and the everyday social and economic realities of the area residents who work in tourism.

A new resort planned for Malgretoute (north side of Petit Piton) will include luxury villas and apartments, along with spas, tennis courts, pools, and a heliport. Given the historical significance of the Pitons in African acts of resistance to colonialism and slavery (there are known sites that were runaway slave camps), it is ironic that the developers named this exclusive luxurious retreat marketed to North Americans and Europeans 'Freedom Bay, St. Lucia.' Freedom as a theme characterizes many tourists' experiences, but resonates with particular strength in Caribbean tourism. The Caribbean is a space to 'free up' oneself from the daily realities of home in a milieu of sun, sea, sand, and sex ensured by unequal power relations of race, class, and gender. In co-opting the Pitons and replacing collective histories of struggle and emancipation with contemporary tourism branding, Freedom Bay embraces and reproduces notions of individual freedom and freedom of markets that neoliberalism imposes on the Caribbean through intensified tourism development.[3]

With regard to these planned resorts, the St. Lucian government – during tenures of both of its political parties – has made some disturbing transactions involving public lands that clearly demonstrate accumulation by dispossession. In the case of Freedom Bay, the government promised to sell public property to the developers, as well as purchased private property for 'public use in tourism' which it then sold to the developers (St. Lucia Government Gazette 2009: 1062). Furthermore, the government and PMA committee agreed to reconfigure the boundaries of a no-build zone to accommodate the developer's plans. At Jalousie, the government is reported to have sold some portions of Queen's Chain, which is land 186 feet away from the high tide mark that is public and may be leased, but not sold (Devaux 2010). Signing over public land to private interests endangers the PMA's status as a World Heritage Site and raises concerns regarding St. Lucia's political and economic sovereignty. The shifting of public and protected land to private interests is also happening outside of the PMA.[4] Furthermore, the creation of exclusive resort spaces, which impedes access to many areas of St. Lucia's World Heritage Site and reserves the most desirable areas for wealthy foreign tourists, challenges the intention and reality of the site's 'universal' description.

In addition to these troubling decisions, some foreign property owners have contravened state development rulings within the PMA, indicating problems of the enforceability of preservation policies and putting the 'universal value' of the site on shaky ground. In a highly publicized case, a wealthy Italian couple obtained an aliens holding license for property within the PMA, but the St. Lucia Development Control Authority (DCA) did not approve their plans to build there. The property owners ignored the DCA decision and began construction, eliciting outraged responses from stakeholders and agencies, such as the Soufriere Marine Management Association and the National Trust. After a legal review, construction was ordered to stop subject to high fines. The Minister of Physical Development was accused of passing blame for the project's approval onto the former government, and his political credibility was called into question in the article 'Does Former Rootsy" Planning Minister Favour the Wealthy?' (Felix 2009).[5]

These developments raised alarms for the World Heritage Committee (WHC). In 2009, the WHC asserted that the PMA's Integrated Development Plan (IDP) was approved without serious consideration for the level of development proposed, noting that 'threats to integrity exist in relation to the protection, planning and management of the property, and the apparent reconsideration of the property's boundaries' (UNESCO World Heritage Committee 2009: 110). The WHC urged reinstating a moratorium on development – in place prior to the IDP – until an assessment of protection and management could be made by the State Party, WHC, and the International Union for Conservation of Nature (IUCN). This moratorium would apply to development plans both inside and outside the PMA boundaries. The PMA committee was unable to satisfy the WHC by the deadline, and the site came dangerously close to being placed on the Heritage in Danger list. The minister accused of lacking the willpower to stop development of the abovementioned villa attended the 2010 WHC meetings and helped persuade the committee not to place the PMA on the Heritage in Danger list. Part of the agreement included restoration of the moratorium on development.

Alternatives within the PMA

Not all heritage tourism efforts in St. Lucia imbricate with dangerous patterns of overdevelopment and exclusivity. There are several sites and events in which the character, interpretation, and goals of blending heritage and tourism are in St. Lucian hands.[6] One of these is the most publicly accessible part of the PMA, the Fond Gens Libres Piton Trail, which offers an alternative model to neoliberal exploitations of heritage tourism.

The trail is a community-based heritage tourism project situated in a small hamlet, Fond Gens Libres (Valley of the Free People). During the period of slavery, the area was alive with Maroon (escaped slaves) and black freedom fighter activity. Until the site was developed, it was known as Unionvale, the estate to which it was once attached. Disassociating from the colonial past through renaming embraces the area's history of resistance and emancipation and reveals a heritage tourism with explicit historical and political implications.

Fond Gens Libres is on the southern side of Gros Piton. The trail is run by part-time guides and co-managed by the Ministry of Forestry and Agriculture. It has long been a popular climbing spot, but the community did not manage the trail or benefit economically from visitors in the past. With assistance from the ministry and the St. Lucia Heritage Tourism Programme, the community developed the site in 2000 and now collects revenue from hikers and is responsible for maintaining the trail.

The site's interpretation center provides historical information on Amerindians and Maroons, as well as oral histories from elder female residents who describe midwifery, medicinal uses for local plants, experiences with poverty, and community life, past and present. The hike includes a stop at the Brigand Cave (one of several unexcavated Maroon sites on Gros Piton), guides naming local plants and animals, and a rewarding view of St. Lucia and the Caribbean Sea from the summit.

However, income from guiding is limited and unpredictable, since most visitors to Soufriere come to view the Pitons not climb them, leading most guides to work in tourism outside of the community. Hikers tend not to stay in the hamlet or to spend much money except to buy refreshments at one of the few rum shops, which are the only businesses in Fond Gens Libres. Despite pressing issues of economic sustainability, Fond Gens Libres represents some of the positive ideals and goals of heritage tourism that UNESCO aims for, but that St. Lucia increasingly finds slipping from its grasp as it is forced to overdevelop for an upscale market.

Importantly, the site is accessible to St. Lucians and is popular with scouts and school groups and during national holidays, such as Emancipation Day. Because of its strong Creole cultural roots, Fond Gens Libres is noted for its Jounen Kwéyòl celebrations, when St. Lucians from around the island visit and participate in heritage festivities. The site is inclusive – available to all visitors – and the community is charged with managing this section of the PMA, which is more in line with the 'universal' spirit World Heritage implies. The very local historical and cultural significance of Fond Gens Libres and Gros Piton – which is, in fact, shaped by very global histories – enhances the meaning and experience of heritage for anyone visiting the PMA.

Conclusion

The Pitons' World Heritage status has, on the one hand, led to the production of a series of policies for land use and development that emphasize preservation, which St. Lucians can reference to oppose or support development projects in the area. On the other hand, the Pitons are now more than ever conceptually and explicitly linked with tourism, marketing, and branding, in ways that are often at odds with St. Lucia's and UNESCO's heritage and preservation objectives. As we have demonstrated here, emphasizing this World Heritage Site as a tourism product has led to agreements between developers and the St. Lucian government to privatize public lands, producing an ever more exclusive social and physical environment in the PMA.

The Caribbean's historical experiences of economic control by external capital interests are perpetuated in the contemporary political economic context that reproduces exploitative colonial relationships of white foreign control over the local black population, culture, and geographic and social space. But while the industry's vulnerability to external forces – hurricanes, recessions, oil prices, market trends – makes tourism jobs insecure, high unemployment rates mean most St. Lucians and indeed the government favor more tourism development for the employment it promises. Increased reliance on tourism throughout the Caribbean should not be understood, however, as the product of local political and economic decision-making, but rather as the consequence of global neoliberal imperatives that have left the region with few alternative pathways for development.

Institutions and NGOs, such as the National Trust and the Soufriere Marine Management Association, are responding to tourism intensification in some heated battles with government representatives.[7] The Trust has expressed serious concerns over 'development pressures on the coastline as well as social implications of declining public access to beaches' (St. Lucia National Trust 2010). More broadly, the St. Lucian public is increasingly speaking out on the airwaves, in print media, and in online forums, in St. Lucia and the diaspora. They object to loss of public space, such as beaches, and to hostile surveillance of St. Lucians on resort beaches, which assures the mobility of a global tourist elite and demobilizes locals in their own spaces of social, cultural, and environmental significance.[8]

That St. Lucians are openly opposed to literally losing ground to neoliberal development agendas is an encouraging sign, as the Jalousie controversy left many dispirited. It might lead to a resurgent activism among St. Lucians working collaboratively with local heritage institutions to challenge socially, culturally, and environmentally irresponsible development. Greater activism at the grassroots level is essential to efforts to make the PMA's World Heritage status more secure and live up to its promise of universal value, which requires not only protecting heritage but also a demonstrated commitment to equal access.

Notes

1 For an examination of the historical role of peasant agriculture in St. Lucian political development, see also Adrien (1996).

2 Intensified tourism development also characterized St. Lucia's preparations for hosting World Cup Cricket in 2007, which followed neoliberal pathways in the form of unique tax advantages for foreign developers who promised to complete projects in time to accommodate the expected influx of cricket fans arriving from outside the Caribbean. Some were not finished in time for the event, but opened sometime after. One of these is the Landings Resort that includes canals and a marina, which required dredging and reconfiguring a beach that was one of the most popular for St. Lucians in the north. The Sapphire Resort, which was to be a large conference hotel, broke ground but never got underway. Construction on Paradis began but remains unfinished due to recession-related problems affecting developers. Thus, some developers benefited from significant tax advantages granted specifically to help St. Lucia meet the demands of hosting an international sporting event, but did not provide the accommodation for that event.

3 Freedom Bay makes this co-opting of symbolically charged physical space of the Pitons explicit on its website: 'This exclusive development is called "Freedom Bay" in commemoration of the slaves of Soufriere who rebelled against the British in 1795. This led to the formation of the Brigands, a freedom movement that eventually liberated St Lucia from imperial rule, albeit briefly. Today we are seeking to create a new freedom by redefining the benchmarks for luxurious living through iconic design and exemplary service.' Available at: www.freedombaystlucia.com/authentic/history/ (accessed July 25, 2016).

4 The government exchanged 228 acres of Crown Lands for shares in a Ritz Carlton development. The status of the resort and the land is uncertain, since the developers used the land as security for a loan from an Icelandic bank that went into receivership as a result of the recent global economic crisis. Similarly, Frigate Island, a private nature reserve on the east coast vested with the St. Lucia National Trust, was signed over by the government to a luxury resort developer. That project ceased, due to the developer's financial difficulties.

5 The article describes how the DCA ordered two food vendors to dismantle a structure they erected to run their business in the capital, Castries, but allowed for the continued construction of a project by wealthy foreigners in violation of DCA rulings. In responses on the newspaper's blog and on radio call-in shows, St. Lucians complained about glaring contradictions in the ways the DCA and other agencies exercise power in situations involving St. Lucians versus non-St. Lucians.

6 Successful community-based heritage projects include the Anse La Raye Fish Fry, a popular Friday night activity with St. Lucians and visitors alike, and the Desbarras Turtle Watch, where St. Lucians and tourists camp on the beach and assist with turtle conservation. Jounen Kwéyòl, a festival exploring and celebrating Creole cultural heritage, involves St. Lucians becoming tourists at home and visiting various communities selected to showcase Creole culture. St. Lucians in the diaspora often make the visit home for this event.

7 Recently the Minister of Tourism, who is a prospective candidate for the Soufriere constituency and, as a former head of Air Jamaica and a resort owner, is connected to global tourism capital networks, engaged in a dispute with the National Trust denying their authority in matters related to tourism development. The Trust publicly reminded the minister of its role in policy formation, as it serves as an advisory committee to government and has a legal mandate for heritage conservation.

8 Exclusionary practices of resorts, like erecting psychological barriers in the form of jetties and beach huts, deter St. Lucians from using beach areas in front of resorts, pushing them to the margins of their own beaches. There is also considerable tension over security procedures on resort beaches (McDonald 2010). Security guards patrol beaches, watching for signs of "locals" who might be "bothering" guests, which can include simple social acts, such as engaging in conversation. St. Lucians know they will be singled out for scrutiny and avoid embarrassment by staying away from resort beaches.

References

Adrien, P. (1996) *Metayage, Capitalism and Peasant Development in St. Lucia, 1840–1957*, Mona, Jamaica: Consortium Graduate School of Social Sciences.

Appadurai, A. and Breckenridge, C. (1991) 'Museums Are Good to Think: Heritage on View in India,' in I. Karp, S. Levine, and T. Ybarra-Frausto (eds.) *Museums and Their Communities: The Politics of Public Culture*, Washington, D.C.: Smithsonian Institution Press.

Devaux, R. (2010) 'A Guide: Cinquante pas du Roi or Queen's Chain,' working paper, St. Lucia: National Trust.

Di Giovine, M. (2009) *The Heritage-scape: UNESCO, World Heritage, and Tourism*, Lanham, MD: Lexington Books.

Felix, S. (2009) 'Does Former "Rootsy" Planning Minister Favour the Wealthy?' *St. Lucia Star*, 9 November. Online, http://stluciastar.com/content/archives/8546 (accessed April 20, 2011).

Freedom Bay St. Lucia. Company Website. Online, www.freedombaystlucia.com/authentic/history/ (accessed April 29, 2011).

Friedman, J. (1983) 'From Plantation to Resort: Tourism and Dependency in a West Indian Island,' unpublished Ph.D. thesis, Rutgers, State University of New Jersey.

Harvey, D. (2007) 'Neo-liberalism as Creative Destruction,' *Annals of the American Academy of Political and Social Science*, 610(1): 21–44.

McDonald, N. (2010) 'Wrong Side of the Beach?' *St. Lucia Star*, July 11. Online, http://stluciastar.com/content/archives/14668 (accessed April 20, 2011).

Reynolds, A. (2003) *The Struggle for Survival: An Historical, Political, and Socioeconomic Perspective of St. Lucia*, Vieux Fort, St. Lucia: Jako Books.

St. Lucia Government Gazette (2009) *St. Lucia Government Printery.* October 19, (42): 1043–1073.

St. Lucia Government Statistics Department (2001) St. Lucia Housing and Population Census Report. Online, www.stats.gov.lc/cenpub_f.pdf (accessed April 12, 2011).

St. Lucia Government Statistics Department (2004) Labour Force Indicators. Online, www.stats.gov.lc/main3.htm (accessed April 12, 2011).

St. Lucia National Trust (2010) Statement to Members and General Public. Online, www.slunatrust.org/memberpages/SLNTStatementtoMembers.pdf (accessed May 10, 2010).

UNESCO World Heritage Committee (2009) 33rd Session Decision. Online, http://whc.UNESCO.org/en/decisions/1831 (accessed April 12, 2011).

Walcott, D. (1992) Litany to the Pitons. *St. Lucia Voice*.
World Travel and Toursim Council (2011) Travel and Tourism Economic Impact, St. Lucia. Online, www.wttc.org/bin/file/original_file/saint_lucia_report_2011-pdf.pdf (accessed April 29, 2011).

6 Archaeological replica vendors and an alternative history of a Mexican heritage site

The case of Monte Albán

Ronda L. Brulotte

'I feel like I am a part of the history here,' said Miguel Chávez, making a sweeping gesture with his arm across the landscape of ruins and, in the distance, his hometown of Arrazola. The 39 year old had spent much of his life traversing the slopes of Monte Albán with his bag of ceramic and stone figures depicting pre-Hispanic deities. It was 9 a.m. and we were among the first to arrive that day at Monte Albán, the southern Mexican state of Oaxaca's most visited archaeological zone and a United Nations Educational, Scientific and Cultural Organization (UNESCO) World Heritage Site since 1987 (see Figure 6.1). In the morning calm, it was difficult to imagine the history of sometimes heated confrontations with employees of the National Institute of Anthropology and History, known by its Spanish acronym INAH. Indeed, the nostalgic inflection in Chávez's voice that morning was later replaced by a tone of indignation:

> They treated us like animals, you know? We just wanted to sell our figures, but they always tried to catch us. If the guards grabbed you, they would take all your pieces and smash them on the ground in front of you. Others would keep your idols and then later sell them and keep the money for themselves.

Chávez described himself as a hard-working individual who was within his legal rights to sell at the site and objected to previous treatment by federal employees. He considered his activities no less legitimate than those of his fellow community members from San Antonio Arrazola, who are internationally famous Oaxacan woodcarvers (Chibnik 2003). But while Arrazola's woodcarvers receive extensive tourism promotion, replica sellers are said to pose a threat to Oaxaca's most important archaeological zone. At issue are claims by some INAH representatives that replica vendors also traffic in original pre-Hispanic artifacts using replica sales as a cover, a claim which vendors denied in their conversations with me. Vendors, in fact, are more commonly criticized for selling their replicas as original archaeological artifacts and constructing elaborate narratives about their provenance and symbolic meaning – in the words of one INAH employee, creating 'a bad image' for Monte Albán. Site staff have been trying unsuccessfully

Figure 6.1 World Heritage Site of Monte Albán. Photograph by the author.

to remove the vendors from Monte Albán since the mid-1970s, but abandoned more aggressive tactics in 1988, after the vendors organized themselves into a legally recognized civil association.

Such disputes remind us that archaeological sites are not neutral places existing outside of the socio-historical conditions of their own modern interpretation; they are often the materialization of class and ethnic politics and the inequalities they represent. Indeed, recent anthropological scholarship rejects reified notions of heritage that squarely lock its significance in the distant historical past (e.g. Abu El-Haj 2002; Breglia 2006; Castañeda 1996; Castañeda and Matthews 2008; Hamilakis and Duke 2007; Mortensen and Hollowell 2009). Interpreting Monte Albán as simply the enduring legacy of Mexico's ancient past elides the process of heritage-making that, I argue, began at the site in the early twentieth century. It is therefore necessary to reexamine paradigms that inherently privilege archaeologists and other technical experts as the rightful interpreters and guardians of cultural materials. I demonstrate how replica artisans and sellers have carved out a niche within the historical development and marketing of the site and, like archaeologists and tourist industry officials, stake a claim to Monte Albán that they regard as no less legitimate. Replica vendors have been active agents in constructing Monte Albán as World Heritage from the beginning. This case study thus provides an alternative telling of the site's history, one which takes vendors' experiences seriously and runs parallel to the story long narrated by archaeology.

Monte Albán is roughly six kilometers east of the capital city of Oaxaca and is considered one of Mexico's most spectacular archaeological remains. The capital of the Zapotecs, Monte Albán was occupied from approximately 500 BC to 750 AD and at its apex was home to 25,000–30,000 inhabitants (Winter 2002). Monte Albán declined around 800 AD but neighboring Mixtec peoples later used the site for religious ceremonies, notably elite burials accompanied by finely worked objects of gold, obsidian, and ceramic. When one such tomb, the now famous Tomb 7, was discovered in 1932, the preeminent Mexican archaeologist Alfonso Caso appraised its contents as the 'richest archeological find in America' (Caso 1932).

Castañeda (1996) illustrates how the reconstructed ruins of another iconic Mesoamerican site, Chichén Itzá in the Yucatán Peninsula, are as much an artifact of early twentieth-century archaeological science as the pre-Hispanic Mayan civilization to which they are credited. His analysis pertains to Monte Albán as well. Modern explorations at Monte Albán first commenced in the nineteenth century, but most of the extensive excavation and reconstruction work was carried out over eighteen field seasons beginning in 1931 under Alfonso Caso. The INAH was created in 1939, with Caso as its first director, and assumed responsibility for Monte Albán and all of Mexico's federal monument zones. Over the next 17 years, INAH spearheaded the transformation of Monte Albán's foliage-covered mounds into the prominent pyramid structures that today dominate the landscape. These massive reconstruction efforts were explicitly framed as advancing scientific knowledge of pre-Hispanic antiquity (Vázquez León 1996) but simultaneously created the conditions for another modern enterprise: the development of Monte Albán as the region's premiere heritage site and tourist attraction.

At the time of my initial fieldwork in 2002–3, and on subsequent trips, replica sellers were an active presence at Monte Albán. They belonged to a legally recognized civil association that counted 60 members. Forty-one men, all but eight from the town of Arrazola, sold ceramic and stone replicas in the site's interior (Figure 6.2). The other 19 vendors, mainly from neighboring communities, offered jewelry to tourists in the parking lot. The replica trade is a difficult livelihood, as vendors' already tenuous existence at Monte Albán is also impacted by the vagaries of the tourism market. For instance, one vendor reported that 'everything went downhill with the twin towers,' referring to the substantial decline in sales after the 9/11 terrorist attacks in the U.S. in 2001. More recently a 2006 teachers' strike culminated in a widespread social movement to oust the state governor. The ensuing civil unrest and violence effectively shut down Oaxacan tourism, which slowly began to recover only to be affected by 2008's global economic downturn.

Vendors enjoyed a modicum of security through their organization but their day-to-day existence at the site depends largely on the good graces of INAH archaeologists and staff. As already mentioned, this relationship remained strained even though it no longer entailed the physical confrontations of earlier periods. The late 1990s saw the implementation of policies intended to curtail the growth of unauthorized vending. Vendors previously performed site maintenance tasks

Figure 6.2 Man selling ceramic and stone replicas in Monte Albán. Photograph by the author.

but were permanently relieved of these duties so that they could no longer use their unpaid labor as a leverage tool with INAH. The vendor association was also required to compose a membership list whose numbers could never be exceeded in the future. Other prohibitions included carrying wares in large duffel bags (only small backpacks are now permitted), eating within the ruins, and setting wares on the ground for display. However, due to the lack of a physical barrier around Monte Albán, INAH was not able to execute its plan to enforce the collection of an admission fee from vendors as they entered.

When I interviewed the then site director, archaeologist Nelly Robles García, in 2003, she backed these policies, though she made clear that she understood the economic dimension of informal commerce. In this spirit, she noted that for years INAH had tried to encourage the vendors to create a legitimate enterprise at the site. They apparently declined offers to designate a space that would serve as an official artisan market, such as the one approved by INAH in 1976 at Mitla (another important Oaxacan archaeological site), which was designed to reduce vendor traffic inside the archaeological zone (Robles García 1996: 269–270). She believed that their refusal stemmed from the fact that many had developed a sales strategy of telling tourists that their homemade replicas were real archaeological artifacts; their stories would not be as effective if they were grouped together in authorized market stalls. She also noted that attempting to sell replicas as originals, if not physically destructive, was morally objectionable.

In making the case against the vendors, Robles cited an incident from 2002 in which a local television crew entered the site with hidden cameras and asked vendors if they had any original artifacts to sell. While most responded that they only sold replicas, three vendors offered pieces they claimed were originals. Tests later proved the pieces to be of recent origin. However, the video segment generated negative publicity for both INAH and the vendors. The vendors subsequently expelled the offenders from the association, but these disciplinary measures did little to sway site officials' opinion of them. For the time being, they begrudgingly tolerate their presence but there is ongoing talk of relocating vendors to a more easily monitored artisan market outside the ruin's main structures.

Modern artifacts

The town of Arrazola, INAH, and Monte Albán are all products of the modern Mexican state, whose power was consolidated after the Mexican Revolution (1910–20). Yet INAH representatives overwhelmingly view the sale of archaeological replicas as an invasive practice that evolved out of economic necessity within a community that, according to one employee, has no 'legitimate historical or cultural connection to Monte Albán.' This perception partly stems from the fact that the vendors' hometown of Arrazola is not a long-standing pre-Hispanic settlement. It was officially recognized as an independent community only in 1937, during Mexico's post-revolutionary land reforms. Prior to that time Arrazola residents were laborers on a small agricultural estate owned by an elite Spanish family. The estate produced sugar cane and attracted rural workers from various parts of Oaxaca, most of whom spoke indigenous languages which eventually gave way to Spanish. While Arrazola residents are descendants of indigenous people, the present day community does not demonstrate geographic continuity or blood descent from a pre-Hispanic settlement in this locality; residents today do not display the obvious linguistic markers of indigeneity. These discourses of place, ethnicity, and identity are themselves complicated issues that I address elsewhere (Brulotte 2012). The question of whether Arrazola should be considered an indigenous community aside, the contemporary power struggle over the replica trade obscures a longer history between the Monte Albán archaeological site and the community.

Arrazola men have always been a dominant force in informal vending at Monte Albán. No official records were kept specifying when it began, but oral histories indicate that the practice originated with on-site excavation field crews, including men from Arrazola, employed by Alfonso Caso in the 1930s. Laborers interacted with tourists and artifact collectors visiting the site, offering them potsherds and other small pieces for a few pesos. Tourism in Oaxaca in the 1930s was limited, with Monte Albán receiving a mere 304 visitors in 1931 (Robles García 1996: 51). The completion of the Pan-American Highway in 1948 opened the region to visitors. An Arrazola resident who worked on Monte Albán excavations in the early 1950s estimated that he was approached two times per week by individuals looking to purchase archaeological fragments. He complied, but maintained that he rarely sold anything larger than ceramic fragments of facial images he found scattered throughout the site. Archaeologists turned the other cheek as long as the workers were not trading off bigger pieces earmarked for private or public collections. International accords guarding against the sale and transfer of pre-Hispanic artifacts were nonexistent at the time.

Inquiries about archaeological pieces may be attributed to shifting trends in the U.S. art market at the time. Pre-Hispanic artifacts were collected and displayed as 'art' for the first time in the United States beginning in the early 1930s (Barnet-Sanchez 1993: 177). Collectors' new-found appreciation of pre-Hispanic art stimulated demand throughout Mexico, creating an extended network of antiquities dealers and the middlemen who supplied them. Often these dealers relied on local people for locating and procuring artifacts, typically men who lived and farmed in close proximity to archaeological zones (García Moreno 2000). Robles García (1996) speculates that Arrazola residents' longtime farming and grazing activities at Monte Albán provided access to artifacts as well as years working on archaeological crews.

Caso's project drew to a close in 1958 but the demand for artifacts continued to grow, and more and more men devoted themselves to the sale of archaeological pieces throughout the 1960s. When original artifacts became difficult to procure, enterprising individuals began to make replicas of the original artifacts to be sold as originals to unsuspecting tourists. The handful of vendors present at the site in the 1950s numbered at least 60 by the early 1970s. One longtime Arrazola vendor who still worked at the site in 2011 described the 1970s as 'a time when almost everybody, all of the men from Arrazola, went to Monte Albán to sell! There was no other way to earn a living back then.'

Several factors precipitated the increase in vending during this period: seasonal work opportunities at Monte Albán had dwindled; the Bracero Program, a guest worker program that imported Mexican laborers to the U.S. during the Second World War (including men from Arrazola), ended in 1964; the woodcarving trade had not yet taken off; and agricultural land was in short supply. Meanwhile, the 1970s was a particularly dynamic period for the Oaxaca tourism industry. The region became a favored destination for European and American backpackers and began to attract Mexican visitors from more affluent areas of the country (Murphy and Stepick 1991). Tourists from the U.S. and Europe increasingly travel to Oaxaca, but even into the new millennium Mexican nationals accounted for nearly 70 percent of the Oaxacan tourism market (SEDETUR 2001).

The founding of the INAH-Oaxaca Regional Center in 1972 paved the way for the tourist development of Oaxaca's archaeological and other cultural attractions. INAH-Oaxaca's mission was primarily anthropological and historical research, but it also created new infrastructure facilitating visits to sites such as Monte Albán. Construction of a visitor center there began in 1973, and the facility was expanded in 1976 with funding from the state tourism office. The complex grew to include a small exhibit area, restaurant, three concession spaces, public restrooms, storage space, and a lobby (Robles García 1996), yet still INAH struggled to accommodate tourist demands.

The surge in tourism stimulated the growth of replica vending, much to the alarm of INAH officials. The legal pretext for INAH to mount its offensive came with the passage of the Federal Law on Archaeological, Artistic, and Historic Monuments in 1972, Mexico's response to the Convention Concerning the Protection of the World Cultural and National Heritage adopted by the General Conference of UNESCO that same year (Litvak et al. 1980). Under the new legislation, all archaeological remnants, immovable structures, and transportable artifacts were the inalienable property of the Mexican nation, and individuals caught removing or tampering with archaeological materials were henceforth in violation of federal law.

The escalating tension between the vendors and INAH after 1972 is documented in a series of written reports submitted from 1975 to 1985 by Monte Albán guards, today archived in the Archaeology Division of the INAH-Oaxaca Regional Center. A recurring complaint concerned the grazing of livestock by Arrazola residents within the boundaries of the archaeological zone; other reports describe the confiscation of original artifacts, which often entailed protracted cat-and-mouse chases and rock-throwing. On at least one occasion, in 1981, an Arrazola man fell and sustained a head wound after being chased by a guard. A letter dated 11 January 1982 documents the center director's request for police intervention at the site.

Arrazola men, too, remembered this period as one filled with daily confrontations that occasionally turned violent. The INAH guards reportedly carried firearms and occasionally used them to scare off the offending vendors, firing shots into the air near them as they fled. Police raids at Monte Albán reportedly employed similar tactics. An Arrazola man, whose ability to outwit INAH guards earned him the nickname the "Rat" from his fellow vendors, recalled what it was like when he first started going to the site in 1976 at the age of 11: 'Sometimes I would be on top of the pyramid and they would chase me down by throwing rocks or firing their pistols. Every day you went up there scared that something was going to happen to you – that you could be arrested or hurt.' Most significantly, what is revealed by the vendor's comment and is corroborated by two INAH reports from 1980 is that by then vendors were no longer exclusively selling archaeological pieces. Rather, faced with increasing pressure from INAH and the state police as well as a dwindling supply of artifacts, the vendors began to offer tourists stone and clay replicas of original pieces. Indeed, what the INAH guards readily confiscated and destroyed as the vendors looked on in dismay were often copies. The same man stated:

> Well, during the time that I went to Monte Albán, that type of thing wasn't going on as much, because it was no longer necessary to sell the old pieces. It was a lot easier to go and buy the replicas already made, treat them [to look old], and then sell them as originals, and in the case that they caught you or something, they couldn't punish you because they were only copies.

This is not to say that vendors had given up selling artifacts entirely. Another vendor estimated that from the late 1970s through the early 1980s his sales comprised 'maybe 10 percent original pieces, mostly caritas, you know, little broken pieces, and maybe 90 percent replicas,' all of which he admittedly sold to tourists as authentic pieces. A memorandum dated 25 November 1980 from INAH archaeologist Marcus Winter to the administrator of the INAH-Oaxaca Regional Center indicates a practice adopted by some vendors of offering a mixed selection of real and imitation artifacts. Inspecting a group of confiscated pieces, he reported that:

> There are five pieces, of which four are stone masks of various colors, green, brown, blue, etc. They were recently made, and therefore they are not Prehispanic. The other piece is a fragment of a grey clay figure, from Monte Albán Period I or II, dating from approximately 200 B.C. to 250 A.D. It is a portion of the figure's face, but is already eroded. This figure is Prehispanic.
>
> (Memorandum, Archaeology Section, INAH-Oaxaca, 1980)

Friction between the vendors and the INAH staff continued into the 1980s, even as the number of vendors began to taper off. The precarious conditions that characterized vending led some to abandon the practice altogether, while those who remained transitioned to selling replica crafts. INAH reframed its discourse against the vendors in terms of their lacking authorization to engage in commercial activities, regardless of the merchandise being offered. In 1985, INAH produced a contract that was distributed to each vendor; by signing this, individuals acknowledged that it was prohibited to sell anything at Monte Albán. Not surprisingly, few were willing to sign the document, and the ones that did paid little attention to it. Vendors always recognized the precariousness of their situation, but their fears of being permanently dislodged from the site were heightened when Monte Albán received its UNESCO World Heritage designation in December 1987.

All of this prompted the vendors to contact a local lawyer, who helped them secure an audience with a representative from the state governor's office. After hearing the case, the official called the director of the Oaxaca police and suspended the use of police force against the vendors. Their lawyer recommended that the vendors form a legally recognized civil association of artisans and craft vendors in order to avert future conflicts. Together they drafted a formal constitution listing the group's by-laws and its 49 founding members and in November 1988 gained federal recognition. The formation of the vendors' association broke the established pattern of physical conflict. Yet the group's new status ultimately proved to be not nearly as important to its day-to-day survival as did the informal agreement established with Monte Albán's director at the time. In exchange for permission to sell their replicas, vendors agreed to perform small maintenance duties, such as picking up trash and cutting weeds, and to watch for vandalism and petty theft. As noted earlier, this arrangement was ended by the next director and today a tense but fragile truce characterizes relations between vendors and site officials.

Competing visions

INAH representatives I interviewed addressed the discrepancy between the celebratory presentation of Monte Albán's ancient history and the largely substandard living conditions of the present-day communities that surround the zone. In the case of Arrazola, they specifically attributed involvement in the replica trade to the community's paltry agricultural land base. However, while these conversations demonstrated that INAH staff were well aware of the conditions underlying contemporary replica vending, few conceded that historical circumstances may have fomented a sense of proprietorship among some Monte Albán vendors that transcended simple economics. My ethnographic data, in fact, revealed more complicated, locally embedded notions of place and ownership. While economic necessity may ultimately trump other factors, most Arrazola residents do not participate in the replica trade due to a lack of education or interest in history – assumptions conveyed by personnel at both the site and the INAH-Oaxaca Regional Center. On the contrary, community members articulate identities and personal histories that are directly tied to Monte Albán.

Arrazola residents have lived and farmed in the vicinity of Monte Albán since well before the first modern archaeological excavations at the turn of the nineteenth century, and then served as a partial labor force for the first 40 years of the site's development. Some of the oldest replica vendors worked on excavations at the site during their lifetimes; others had fathers and grandfathers who assisted on early projects. On one occasion, an elderly vendor showed me his family's most cherished pieces of memorabilia: a photograph taken sometime in the 1930s of his grandfather guiding an expedition to Monte Albán. Although the quality and framing of the photo were poor, the vendor assured me that one of the men was none other than Alfonso Caso. The man then pointed out details about each individual in the party, adding his own interpretations of that day's plausible events. He keenly noted that his grandfather, dressed in the homemade clothes of a farmer, stood apart from the finely dressed group of men and women, and that he appeared to be much darker, 'more Indian,' than the others. The personal connection the man felt with this piece of Monte Albán's history was evident in both his reverent narration of the photograph and the care the family accorded it.

The same man proudly recounted how his grandfather later found and helped unearth Monte Albán's famous Tomb 7, one of the most spectacular moments in modern archaeological history. His story is corroborated by a synopsis of Tomb 7's discovery in an encyclopedia of Oaxacan history (Zárate Aquino 1995). This detail is glossed over in the now commonplace narrative of Tomb 7's 'discovery' by Caso and his team of archaeologists. To cite another case, the man who is widely regarded as Arrazola's best replica artisan has a law degree from a local university and is well versed in Oaxacan politics, history, and economics. In researching designs for his pieces, he has accumulated an impressive knowledge of pre-Hispanic iconography and history. We should not be surprised by the sense of proprietorship expressed by vendors, who see themselves as having a more intimate knowledge of Monte Albán than many of the outsiders who work at the site; its terrain constitutes part of the everyday physical and ideological landscape inhabited by Arrazola residents.

Chalking up the replica trade to a lack of education or inadequate land base obscures its place within a more complex set of historical relationships that emerged after the Mexican Revolution and developed during subsequent decades of the twentieth century: the advent of modern archaeological science, an expanding global tourism industry, the rise of international markets for Mexican art and artifacts, migration, and a nationalist project enacted through pre-Hispanic cultural patrimony. From this perspective, INAH, Monte Albán, archaeologists, replica sellers, and the town of Arrazola itself are all part of a modern 'heritage-scape' (Di Giovine 2009) constructed through the apparatus of the modern Mexican state. Institutional discourses of replica vending at Monte Albán that depict sellers as invading an otherwise pristine archaeological site elide the specificities of these historical processes The history of the replica trade at Monte Albán is not incidental to but rather an integral part of the site's modern history.

References

Abu El-Haj, N. (2002) *Facts on the Ground: Archaeological Practice and Territorial Self-Fashioning in Israeli Society*, Chicago, IL: University of Chicago Press.

Barnet-Sanchez, H. (1993) 'The Necessity of Pre-Columbian Art in the United States: Appropriations and Transformations of Heritage, 1933–1945', in Boone, E.H. (ed.) *Collecting the Pre-Columbian Past*, Washington, DC: Dumbarton Oaks Research and Library Collection.

Breglia, L.C. (2006) *Monumental Ambivalence: The Politics of Heritage*, Austin: University of Texas Press.

Brulotte, R.L. (2012) *Between Art and Artifact: Archaeological Replicas and Cultural Reproduction in Oaxaca, Mexico*, Austin: University of Texas Press.

Caso, A. (1932) 'Monte Albán, Richest Archaeological Find in America', *National Geographic*, (62) 4: 487–512.

Castañeda, Q.E. (1996) *In the Museum of Maya Culture: Touring Chichén Itzá*, Minneapolis: University of Minnesota Press.

Castañeda, Q.E. and Matthews, C.N. (eds.) (2008) *Ethnographic Archaeologies: Reflections on Stakeholders and Archaeological Practices*, Lanham, MD: AltaMira Press.

Chibnik, M. (2003) *Crafting Tradition: The Making and Marketing of Oaxacan Wood Carvings*, Austin: University of Texas Press.

Di Giovine, M.A. (2009) *The Heritage-scape: UNESCO, World Heritage, and Tourism*, Lanham, MD: Lexington Books.

García Moreno, G.Z. (2000) 'Guardianes y moneros: Patrimonio arqueológico y supervivencia campesina en el sur de Nayarit', unpublished M.A. thesis, Centro de Investigaciones y Estudios Superiores en Antropología Social de Occidente, Jalisco, Mexico.

Hamilakis, Y. and Duke, P. (eds.) (2007) *Archaeology and Capitalism: From Ethics to Politics*, Walnut Creek, CA: Left Coast Press.

Litvak King, J., González R.L. and González, M. (eds.) (1980) *Arqueología y Derecho en México*, Mexico City: UNAM.

Mortensen, L. and Hollowell, J. (eds.) (2009) *Ethnographies and Archaeologies: Iterations of the Past*, Gainesville: University of Florida Press.

Murphy, A.D. and Stepick, A. (1991) *Social Inequality in Oaxaca: A History of Resistance and Change*, Philadelphia, PA: Temple University Press.

Robles García, N.M. (1996) 'El Manejo de los Recursos Arqueológicos en México: El Caso de Oaxaca', unpublished Ph.D. thesis, University of Georgia, Athens.

SEDETUR (Secretaría de Desarollo Turística del Estado de Oaxaca) (2001) *Informe Encuesta Perfil del Visitante a la Ciudad de Oaxaca*, Oaxaca: SEDETUR.

Vázquez León, L. (1996) *El Leviatán Arqueológico: Antropología de una Tradición Científica en México*, Leiden, Netherlands: Research School, CNWS.

Winter, M.C. (2002 [1989]) *Oaxaca: The Archaeological Record*, Mexico City: Editorial Minutiae Mexicana.

Zárate Aquino, M. (1995) *Pequeño Diccionario Enciclopédico de Oaxaca*, Oaxaca: Universidad José Vasconcelos de Oaxaca.

7 Indigenous perspectives on ownership and management of Yucatecan archaeological sites

Stephanie J. Litka

This chapter examines the complex and multifaceted nature in which Maya tourist workers conceptualize control and ownership over archaeological tourism in the Yucatan Peninsula. By highlighting how local employees mediate between tourist, government, and private business expectations of ideological, monetary, and spatial management, I show how the Maya strategically utilize these tensions to construct partible identities (Strathern 1996) on both community and individual levels. Hence, this idea expands on the traditional view of ownership as a dichotomous 'us' versus 'them' relation of local resistance. Rather, I take a political-economic approach to understanding how the Maya view identity as a process that depends on these tensions as opposed to an entity that exists as a mere reaction to tourism. As local populations are becoming increasingly globalized, many inhabitants demonstrate dynamic linguistic and cultural traits that both conform to external expectations and adhere to local values of host/guest relations.

I specifically focus on how Chichen Itza serves as a default site for understanding how power relations are manifested and negotiated in the nearby archaeological community of Coba. This town has seen an increase in tourism over the past few decades, and local employees now depend on this industry as part of their economic livelihood. Although Coba differs from Chichen Itza in terms of historical timeframe and cultural influences, the latter tends to overshadow these distinctions by remaining one of the most prominent symbols of the ancient Maya past. Here I examine how the international notoriety of this site brings forth issues of ideological, monetary, and spatial control among differing parties in Coba. Rather than view external notions of ownership and management in opposition to local political economy, I examine how the Maya utilize these tensions as a way of constructing a multifaceted sense of identity in their daily lives. I begin with a theoretical introduction to tourism as a means of understanding how each element of control fits in with the broader issues of local/global relations that arise from the tourist industry. I then examine individual issues of ideological, monetary, and spatial power, linking case studies in both locations with ethnographic research to demonstrate how Coba employees reconcile these tensions for their benefit. I conclude by proposing avenues of sustainable tourism that address the needs of local community members while taking into account the expectations of government policy, private businesses, and tourists for future development.

Background of tourism in the Yucatan Peninsula

Before its rise as one of the biggest tourist destinations in the world, Cancún was a small village with about 400 inhabitants. It wasn't until the creation of FONATUR (the state tourism ministry in Mexico) in the 1970s that this industry expanded tremendously under the presidency of Echevarria (Pi-Sunyer et al. 1999). Its location near prominent archaeological sites and the Caribbean coast made it an ideal location to develop for heritage, recreational, and eco tourism. The Maya in particular are indigenous peoples living throughout Mesoamerica, in the nations of Mexico, Guatemala, Belize, and parts of Honduras. They speak over 30 distinct languages (Campbell and Kaufman 1985) and maintain cultural practices that reflect their connection to the ancient past. Their historical achievements in astronomy, mathematics, and architecture continue to receive worldwide attention among scholars and tourists today. In the context of national promotion of these zones, various Mayan communities are confronting globalization directly.

Both Chichen Itza and Coba are located within a few hours of Cancún and Merida, the capital city of the state of Yucatan. Their easy access via paved roads and the availability of first-class public transportation enable tourists to visit each site and interact with indigenous guides, waiters, shopkeepers, and hotel workers during their stay. Located in the state of Yucatan, Chichen Itza stands as one of the foremost symbols of Mayan culture. Through the analysis of radiocarbon and long-count dates, Andrews and colleagues (2003: 152) suggest that this city emerged as a regional capital during the late Classic period, fell into decline, and then re-emerged before its final decline in 1224 AD. Covered in jungle for over 700 years, formal excavations began in the 1920s and it was declared a United Nations Educational, Scientific and Cultural Organization (UNESCO) World Heritage Site in 1987 (Breglia 2005: 386). The town of Coba is located in the state of Quintana Roo, one hour from the Riviera Maya, and is also adjacent to the archaeological site of its namesake. Historically, it was one of the largest centers of ancient Mayan culture, reaching its pinnacle during the Classic Maya period between 600 and 900 AD (Coe 1999). Today, it boasts of having the highest pyramid in the Yucatan Peninsula, serving as one of the few sites that tourists can still climb in the region. It has witnessed a sharp increase in international tourism over the past few decades, and nearly half of the working population depend on this industry for their economic livelihood (Walker 2005). At the same time, there remain several indigenous elements in Coba, including language use, dress, and social ties, due to its smaller size and semi-remote location from the Riviera Maya coast.

Theoretical perspectives

Through an analysis of authenticity, narrative, goods, and performance relating to ideological, monetary, and spatial control, I provide a theoretical framework for understanding indigenous tourist management. Over the past 30 years, the study of tourism has grown extensively as an academic field through its interdisciplinary approach to understanding host and guest relations. Various debates have specifically centered around the idea of tourism and its effects on local communities. According to van den Berghe (1994: 8), the notion of 'the quest for the exotic other' leads visitors to seek native populations that are 'in as untouched, pristine, and authentic form as can be found.' MacCannell (1992) discusses the problematic nature of tourism's propensity to dichotomize groups in terms of primitive/modern, uncivilized/civilized, and folk/urban. These oppositions contribute to preconceived notions about the 'Other' and often disregard the various factors that have modified their culture throughout space and time. This can lead to tensions with respect to how indigenous workers address these expectations while simultaneously striving for control through their own participation in this industry.

García Canclini (1995) and Appadurai (1996) offer a more nuanced approach to indigenous conceptions of globalization by suggesting that local workers become 'cultural hybrids' through the flow of peoples and ideas associated with this phenomenon. While this may seem like a recent phenomenon, a closer look at historical processes reveals how indigenous populations have incorporated elements of cultural 'Others' into previous systems of social and economic exchange. Even among Maya societies themselves, Joyce (2005) notes how they have had contact with other groups and spoken different languages throughout their history, not just in the modern era. In a related manner, Chambers (2000: 98) questions the notion of 'authenticity' in the first place, arguing that the past is no more real than the present. Hence, his understanding of this term is that it occurs when locals have 'significant control over their affairs, to the extent that they are able to play an active role in determining how changes occur in their social settings.' Handler and Saxon (1998) support this idea by asserting that the reconstruction of tradition has characterized human culture just as much in the past as in the present day.

On the other hand, Chambers (2000: 99) highlights how local producers commonly take advantage of stereotypical portrayals of their culture for economic, social, and political benefits. Ann Tsing's (2005) notion of friction offers a more nuanced approach to understanding the dynamics of host/guest relations with the idea that unequal encounters and expectations can lead to new arrangements of power. Accordingly, she notes how 'cultures are continually co-produced through frictions, which are the awkward, unequal, unstable, and creative qualities of interconnection across difference' (p. 4). Among Yucatec Maya workers themselves, Castañeda (1996: 9) adds that they participate in the invention of culture imposed by outsiders as a means of becoming active agents to both engage with and contest Western knowledge through their involvement in the tourism industry. By utilizing the multiparty tensions involved in the tourist realm, indigenous employees strategically adhere to outside expectations which enable them to participate in the global market economy and to become agents of their own cultural change. It is through these tensions, therefore, that issues of control and ownership are constantly mediated through the interdependence of all parties involved.

Regarding ideology and performance, Bendix (2002: 472) theorizes that they are driven by both actual and desired experience on the part of hosts and guests. Chambers (2000: 113) adds that the demands of tourism often encourage guides and other workers to reenact ceremonies that either may not persist in the daily lives of contemporary populations or occur out of cycle to cater to visitors' schedules. In such cases, the social value placed on them by natives can diminish over time, as they are packaged for the benefit of the tourist industry. Such performances often correspond to the notion of 'invention of tradition' (Ranger and Hobsbawm 1992), where the origins of the practices may be much more recent than the audience realizes. From the tourists' perspective, MacCannell (1992) feels that many of the performances serve as a means of escaping an alienated reality in the hope of finding a more pure form of social authenticity that existed before the modern era. He argues that while this goal is never fully tangible, the desire of visitors to escape into this 'nostalgic' form of reality drives them to witness various types of sites, narrations, and shows that are enacted for their entertainment.

According to Muriel (2001: 56–57), there are between 200,000 and 250,000 archaeological sites throughout Mexico. Some are small and remain covered in jungle, while others (such as Chichen Itza) are enormous ancient metropolises that have been highly excavated using public money. Early archaeologists tended to focus on reconstructing the largest pyramids, thereby emphasizing how the ancient elite lived for public benefit, while overlooking the general population as a whole (McClung de Tapia 2002). The New Archaeology Movement in the 1970s (Muriel 2001: 57–58), however, focused more attention on overall settlement patterns and cultural ecology rather than sheer restoration. Theoretical debate continues today between scholars who highlight the importance of the restoration and public presentation of World Heritage Sites and those advocating further scientific research. Walker (2005) notes how these schisms center around issues of which buildings are to be restored, which time period is to be emphasized, for whom the ruins are to be reconstructed, and how much money to invest in their overall preservation.

Various scholars have discussed issues of space with respect to power relations and tourism development. Zukin (1991) describes the creation of 'landscapes of power' in which spaces and places are contested by hosts, guests, and external agencies alike. Breglia's (2005) notion of 'monumental ambivalence' explains how cultural patrimony is constantly negotiated by all parties involved. For example, she observes how the spatial proximity of archaeological sites to indigenous communities serves as a means for the latter to claim ownership over them. This is not necessarily attributed to a personal attachment to their ancient ancestors, but has more to do with jobs that are passed down from fathers to sons in certain villages located adjacent to these zones. On a similar note, other scholars, such as Greathouse-Amador (2005) and Timothy and White (1999) have argued that the closeness to monuments provides an incentive for local towns to reclaim their heritage by learning more about their own history in order to convey this information from a local point of view.

Case studies of power relations in Chichen Itza and Coba

The Mexican government asserts its ideological control by advertising Chichen Itza as one of the foremost archaeological symbols of the Mayan world. While other sites are equally significant among the ancient Maya, they tend to be overshadowed by Chichen Itza's notoriety. The famous El Castillo monument draws tourists from all over the world to witness its archaeological grandeur, especially during the spring and fall equinoxes, when the image of a descending snake appears on the pyramidal side. Ironically, the origin of the Feathered Serpent came from Central Mexican cultures, even though the Maya did incorporate the symbol into their religious pantheon under the name of Kukulkan (Read and Gonzalez 2002). Nevertheless, the characteristics associated with this Maya–Toltec site, such as architectural style and religion, are popularized through mainstream advertisements, elaborate performances, and light shows. Spatial and temporal variations from other sites tend to be minimized, allowing Chichen Itza to serve as an all encompassing representation of Mayan culture. As Alarcón (1997: 182) notes, this site 'may be experienced simultaneously as pleasure, education, work, adventure, and time travel.' This type of accommodation to the needs of large numbers of tourists, however, does not extend to other locations that receive less attention on an international scale.

Large sums of money are poured into excavating Chichen Itza due to its worldwide notoriety. Ever since Chichen Itza was declared a UNESCO World Heritage Site (and more recently one of the New Wonders of the World), the Mexican government has invested heavily in this location. In comparison, only 5% of Coba has been excavated, leaving most of the site covered in jungle. Taube (2001) notes that only certain parts of this site are made accessible to the public in order to reduce maintenance costs, as most of the buildings are covered under vegetation. As a result, there is little to see at the site itself, which means that the tourists stay for a few hours and then leave without partaking in other activities (such as movies and light shows) that could provide more entertainment for them and jobs for the community. In a related manner, tour groups often follow strict schedules and allow limited time to see the site, leaving many tourists feeling rushed and not able to fully absorb Coba's history. The monetary dominance of Chichen Itza extends to goods and souvenirs, as several replicas of the famous El Castillo monument are found in Coba's shops. On the other hand, tourists have a harder time finding those of Nojoch Mul, despite the fact that it is located in Coba and is one of the tallest ruins in the Yucatan Peninsula.

With respect to the spatial dominance of Chichen Itza, this area is highly developed and commercialized beyond the physical monuments themselves. On the other hand, the ambience surrounding Coba is much less so, as the local culture is more apparent in this location. The employees who work inside the property of Chichen Itza tend to have formal business and linguistic training, catering to the needs of international tourists who visit the site daily. While there is more of an indigenous presence outside of the archaeological zone, this remains marginalized compared to the more official-looking stores inside the site. Breglia

(2005) notes how even INAH-trained 'custodios' who used to live amidst the ruins have recently been relocated to the nearby town of Piste in order to modernize the site. This has resulted in a shift of labor, social, and kinship relations that continue to be negotiated among government, private sector, and local workers who claim a stake in the ownership of Chichen Itza. Even though Coba maintains a more natural and locally owned ambience, increased urbanization along the Riviera Maya may entail further development of the site in order to adhere to both INAH's and tourists' expectations of appearance and service.

Linking power with ethnography

Over the past several years, I have conducted ethnographic research among tourist workers in Coba. Through participant observation, interviews, taped conversations, and guided tours with over 50 local employees, I examined the ideological impact of Chichen Itza in their daily lives. One observation was that visitors tended to view the Maya as a homogeneous group over space and time. Many tourists used Chichen Itza as a default for understanding ancient beliefs and practices in Coba. Questions such as 'Was this done in Chichen Itza?' and 'Why is this different from there?' occurred frequently among them. Others expressed disgust over sacrifices and warfare tactics found at the latter location, in addition to using Chichen Itza as a basis for comparing architectural styles, leading to preconceived notions of Coba upon visiting the site.

Nevertheless, local employees (especially tour guides) often used these tensions for their benefit when conveying Coba's history to others. The frictions revolving around Chichen Itza's ideological control enabled workers to learn more about their own past by understanding Coba's relation to the broader Mayan civilization. When interviewed, many of my research participants felt that an important characteristic of being 'Maya' meant knowing something about their heritage. This was especially true with respect to understanding Coba's spatial and temporal relation to other well-known archaeological sites. On almost every tour, locals distinguished between the 'puro Maya' culture of Coba and the 'Maya Tolteca' culture of Chichen Itza. The former site was built much earlier, sharing many stylistic and religious features with the sites of other Mayan groups in Guatemala rather than in Central Mexico.

Another element of ideological control pertains to the portrayal of ancient sacrificial events. Nearly all tour guides emphasized how Coba's inhabitants practiced blood-letting for religious purposes, whereas in Chichen Itza full human sacrifices were conducted for both religious and warfare events. One guide was quoted as stating that Chichen Itza 'ya es otra cultura, de influencia tolteca, no es maya 100%' (it is another culture, of Toltec influence, it's not 100% Maya). In this manner, the modern Maya strive to clarify misconceptions and generalizations (and actually depend on these frictions) that arise when tourists visit their communities. They do this by utilizing Chichen Itza's popularity to highlight stylistic and symbolic differences found there and at other sites throughout the region. Hence, this site serves as a focal point for locals and tourists alike to better understand the spatial and temporal distinctions of ancient Mayan civilization.

All of the participants interviewed were also enthusiastic about sharing their contemporary culture with visitors, especially since many tourists believed that the Maya had disappeared when the ruins were abandoned. One woman stated in an interview, translated from Spanish, that 'we are still here and we want them [the tourists] to know about us.' While they acknowledged that the Spanish Conquest had drastically altered their religious and ideological practices over the centuries, many were quick to point out that distinctions have always existed between the ancient Maya city-states throughout the region. As opposed to examining control from a purely external perspective, therefore, local narrations serve to increase dialogue with visitors as a means of promoting greater cultural awareness through the tourism industry.

In terms of monetary control, many local employees were in favour of future excavations and research in Coba, both for personal knowledge of their history and for the sake of tourism. At the same time, they realized that Coba was developed enough to attract visitors while also maintaining a jungle-like appeal for a wide variety of audiences. Hence, they emphasized the natural appeal of their ruins that had remained reasonably 'untouched,' catering to the tourists' desire to experience a more 'authentic' view of how the Maya had lived in the past. In several interviews, workers expressed hesitation over developing Coba to the same extent as Chichen Itza and other highly commercialized sites. They wanted this site to retain its distinctiveness in order to serve the interests of hosts and guests alike. Furthermore, Nojoch Mul is one of the few monuments that visitors are able to climb in the Mayan region. Local *tricitaxista* services that offer to take tourists to and from this ruin provide an extra source of income for young men, since it is situated over two kilometers from the site entrance. Indigenous workers, therefore, strategically utilize this situation to advocate further development while taking advantage of tourists' desires for economic and ecological benefits.

With respect to spatial control, previous research shows that while INAH controls the ruins of Coba, the town rests on ejido land that still remains a powerful social and political entity today. Not only do the inhabitants maintain local authority, but they actually utilize their close location to the ruins to share their culture with tourists. Both tour agencies and locals alike advertise this town as a Mayan village. Many companies offer packages that include tours to the ruins and to selected homes in order for visitors to see how the modern Maya live. Several workers in my study reported that visitors wanted to witness 'traditional' practices (such as making tortillas by hand and weaving hammocks) in order to gain a more authentic insight into the daily life of indigenous communities. They often took pictures and spent time exploring all parts of a household as part of a brief detour after exploring the adjacent ruins. The host population is able to benefit monetarily from visitors and gain social status through increased communication with foreign guests who want to learn more about them.

Sustainable tourism

Having addressed how local employees conceptualize ideological, monetary, and spatial control over tourism management, this chapter now proposes some avenues of sustainable tourist development that meet the interests of each party involved. The first one entails increased communication between locals, private businesses, government officials, and tourists to promote greater cultural sensitivity and awareness. For example, Ardren (2002) observes how joint collaborations between archaeologists and indigenous citizens allow the latter to take a more active role in conveying their contemporary culture with reference to the ruins that symbolize their ancient past. This mutual dependency would serve to 'reverse the power imbalance present in Western academic control of Maya culture' (380) by taking into account how local people's perceptions of their history are conveyed to visitors at each site.

Secondly, increased business and linguistic training is vital in order for indigenous workers to assume control over local tourism management. Although tour guides are required to become certified by INAH to work at archaeological sites, other employees such as shopkeepers, waiters, and hotel staff often lack the formal commerce skills needed to actively meet the needs and interests of multiparty visitors on a long-term basis. Hosts and guests alike realize the importance of this training if tourism is to remain under local control in the future. Some examples are greater fluency in other languages, more efficient customer service, and better accessibility of roads between different points within Coba. By complying with both national and international standards of employment, native Yucatec speakers would have the skills to retain their language and culture while gaining the knowledge and experience needed to successfully manage tourism in the years to come. While there are several other avenues of sustainable tourism, increased dialogue with and better training methods for local workers provide a foundation for better understanding how the interests of all parties can be met in the future.

I have examined here the various ways in which Maya tourist workers conceptualize ownership and mediate between government, tour agencies, private businesses, and visitors' expectations of community-based tourism. By focusing on how Chichen Itza serves as a default location for understanding power relations in the archaeological zone of Coba, I demonstrated how Yucatec Maya employees reconcile issues of ideological, monetary, and spatial control through a constant negotiation of the tensions and frictions involved in this industry. While indigenous employees in Coba acknowledge Chichen Itza's touristic dominance in the region, they also utilize its power for their own advantage in several ways. Moreover, local workers actually depend on multiparty expectations of tourist management (often based around Chichen Itza's standards of operation) to simultaneously become active participants in the global market economy and maintain social ties within their community. In turn, these tensions help shape a dynamic sense of identity that is constantly shifting among locals in Coba. Taking into account the ways in which the indigenous inhabitants of Coba both reconcile and depend on tourism in their daily lives, this chapter ultimately aims to promote sustainable tourism for all parties involved in the years to come.

References

Alarcón, D.C. (1997) *The Aztec Palimpsest Mexico in Modern Imagination*, Tucson: University of Arizona Press.

Andrews, A.P., Andrews, E.W. and Castellanos, F.R. (2003) 'The Northern Maya Collapse and Its Aftermath', *Ancient Mesoamerica*, 14 (1): 151–156.

Appadurai, A. (1996) *Modernity at Large: Cultural Dimensions of Globalization*, Minneapolis: University of Minnesota Press.

Ardren, T. (2002) 'Conversations about the Production of Archaeological Knowledge and Community Museums at Chunchucmil and Kochol, Yucatán, Mexico', *World Archaeology*, 34 (2): 379–400.

Bendix, R. (2002) 'Capitalizing on Memories Past, Present, and Future: Observations on the Intertwining of Tourism and Narration', *Anthropological Theory*, 2 (4): 469–487.

Breglia, L.C. (2005) 'Keeping World Heritage in the Family: A Geneology of Maya Labour at Chichen Itza', *International Journal of Heritage Studies*, 11 (5): 385–398.

Campbell, L. and Kaufman, T. (1985) 'Mayan Linguistics: Where Are We Now?', *Annual Review of Anthropology*, 14: 187–198.

Castañeda, Q. (1996) *In the Museum of Maya Culture*, Minneapolis: University of Minnesota Press.

Chambers, E. (2000) *Native Tours: The Anthropology of Travel and Tourism*, Long Grove, IL: Waveland Press.

Coe, M. (1999) *The Maya*, 6th edition, New York: Thames and Hudson.

García Canclini, N. (1995) *Hybrid Cultures: Strategies for Entering and Leaving Modernity*, Minneapolis: University of Minnesota Press.

Greathouse-Amador, L.M. (2005) 'Tourism: A Facilitator of Social Awareness in an Indigenous Mexican Community?', *Review of Policy Research*, 22 (5): 709–720.

Handler, R. and Saxon, W. (1998) 'Dissimulation, Reflexivity, Narrative, and the Quest for Authenticity in "Living History"', *Cultural Anthropology*, 3 (3): 242–260.

Joyce, R.A. (2005) 'What Kind of Subject of Study is "The Ancient Maya"?', *Reviews in Anthropology*, 34: 295–311.

MacCannell, D. (1992) *Empty Meeting Grounds: The Tourist Papers*, London: Routledge.

McClung de Tapia, E. (2002) 'Exchanges: A First Look at Public Outreach in Mexican and Guatemalan Archaeology', *SAA Archaeological Record*, 2: 27–29.

Muriel, A.M. (2001) 'Archaeological Research in Mexico's Monumental Sites', in R.D. Drennan and S. Mora (eds.) *Archaeological Research and Heritage Preservation in the Americas*, Washington, D.C.: Society for American Archaeology.

Pi-Sunyer, O., Thomas, R.B. and Daltabuit, M. (1999) *Tourism and Mayan Society in Quintana Roo, Mexico*, Storrs, CT: Center for Latin American and Caribbean Studies at the University of Connecticut.

Ranger, T. and Hobsbawm, E.J. (eds.) (1992) *The Invention of Tradition*, Cambridge: Cambridge University Press.

Read, K.A. and Gonzalez, J.J. (2002) *Handbook of Mesoamerican Mythology: A Guide to the Gods, Heroes, Rituals, and Beliefs of Mexico and Central America*, Oxford: Oxford University Press.

Strathern, A. (1996) 'Seeking Personhood: Anthropological Accounts and Local Concepts in Mount Hagen, Papua New Guinea', *Oceania*, 68 (3): 170–188.

Taube, K. (2001) 'Ancient and Contemporary Maya Conceptions of the Field and Forest', Paper Presented at the 21st Symposium in Plant Biology, University of California, Riverside.

Timothy, D.J. and White, K. (1999) 'Community-Based Ecotourism Development on the Periphery of Belize', *Current Issues in Tourism*, 2 (2–3): 226–242.

Tsing, A.L. (2005) *Friction: An Ethnography of Global Connection*, Princeton, NJ: Princeton University Press.

van den Berghe, P.L. (1994) 'Marketing Mayas: Ethnic Tourism Promotion in Mexico', *Annuals of Tourism Research*, 22 (3): 568–588.

Walker, C. (2005) 'Archaeological Tourism: Looking for Answers along Mexico's Maya Riviera', *National Association for the Practice of Anthropology Bulletin*, 23 (1): 60–76.

Zukin, S. (1991) *Landscapes of Power: From Detroit to Disney World*, Berkeley: University of California Press.

8 World Heritage, tourism development, and identity politics at the Tsodilo Hills

Rachel F. Giraudo

When the United Nations Educational, Scientific and Cultural Organization (UNESCO) adopted the Convention Concerning the Protection of World Cultural and Natural Heritage in 1972, it also helped to launch the study of a number of theoretical and practical issues related to World Heritage. They include, among others, the very definition of 'heritage' (e.g., Smith 2006), site conservation methods (e.g., Leask and Fyall 2006), and the supposed 'outstanding universal value' that World Heritage Sites possess (e.g., Titchen 1996; Cleere 2001). Examination of these issues subsequently led to policy revisions and prompted further scholarly investigations into World Heritage as an emerging global phenomenon. As recent publications (e.g., Di Giovine 2008; Harrison and Hitchcock 2005) and the other chapters in this volume demonstrate, tourism is an important and timely topic in both the scholarship and the practical management of World Heritage Sites. However, studies of World Heritage and tourism generally focus on conservation and management issues (e.g., Pedersen 2002). Social effects, such as the ethnic and gender identity politics of nearby communities, are sometimes overlooked. Studies that do consider identity politics reveal that World Heritage status does have a significant influence on local communities, usually through the culturally commodifying tendencies of tourism that accompany the status (e.g., Tucker 2003; Owens 2002).

The effects of World Heritage and tourism on identity politics are most striking at contested heritage sites and those located in developing countries. Contested heritage sites are already politically charged between various stakeholders, so World Heritage status, conservation protocols, and the ensuing economic development of these properties all intensify existing social problems or create new ones. World Heritage status can exacerbate existing tensions among different ethnic, religious, and/or political groups who lay claim to the same heritage site and vie for their group's recognition and control of the site. Likewise, developing countries pursue World Heritage status not only to assist with heritage conservation, but also because they want to deploy the World Heritage brand at their heritage sites in order to increase tourism and provide development opportunities to those living nearby. The hope of these countries and the non-governmental organizations (NGOs) and other development agencies working with them is that tourism to World Heritage Sites will help alleviate poverty for local communities.

Community-based tourism is meant to reverse the top-down power dynamic of external agencies setting the development agenda by attempting to put communities in charge of their own development (Fabricius et al. 2004). However, this approach is not always successful, in part because there are other social issues that need to be considered (see Reed 1997). Heterogeneous, multi-ethnic communities have complicated power relations that need to be assessed to adequately comprehend what the 'community' entails, as it is the 'community' as a population unit that development facilitators envision in the planning of their initiatives. This concept potentially blurs ethnic, gender, and class divisions as it is commonly used as a geographical reference and not as a more nuanced and sociopolitical reference (see Geschiere 2009: 77, 81–83).[1] Additionally, studies show that development initiatives reflect the gender norms of the agencies that create them, and that the implementation of these initiatives affects the gender politics of the community members on which they are imposed (e.g., Schroeder 1999). Community-based development desperately requires careful attention to be paid to the situational dynamics of each community in which such projects are proposed.

It is clear that World Heritage status affects, to some degree, the identity politics of community members living near World Heritage Sites, especially those sites inscribed for cultural heritage. This is due to significantly increased tourism, cultural commodification, and development initiatives, all of which can alter traditional ethnicity and gender roles. Thus, World Heritage status does not actually entail cultural conservation, but is instead an instigator of cultural transformation through both tourism and development initiatives that overwhelmingly come along with the status in developing countries. Anthropological studies of tourism and cultural commodification generally focus on the question of authenticity and less often examine the topic as part of a global economic process (but see Kirshenblatt-Gimblett 2006). However, by framing the problem of the identity politics of World Heritage and tourism through the development lens, it is possible to also investigate the sociopolitical issues that arise.

Instead of working in the abstract or through generalization, engaged ethnographic fieldwork greatly contributes to scholarship and informs policies on the entanglements of World Heritage and tourism. These 'heritage ethnographies' highlight the gradations of meaning in the case studies investigated, and can overturn prevailing understandings of the social effects of World Heritage and tourism (e.g., Breglia 2006; Fontein 2006; de Jong and Rowlands 2007). By ethnographically grounding academic and policy claims with extensive fieldwork in particular locales, and not just through consultation and survey, we gain a more enriched understanding about the global phenomenon of World Heritage and its impact in economically developing parts of the world. This enhanced awareness is important as new State Parties ratify the 1972 World Heritage Convention to promote the inclusion of their heritage sites on the World Heritage List and, in particular, in light of the World Heritage Centre's agenda to have countries in the developing world list more World Heritage Sites (e.g., African World Heritage Fund).

As part of my heritage ethnography, I explore issues of ethnic and gender identity politics at the Tsodilo World Heritage Site in northwestern Botswana. My analysis, based on more than two years of ethnographic fieldwork and archival research (2007–2009), examines the implementation of a new heritage management plan for the site that aims to improve the livelihoods of the nearby Ju/'hoan and Hambukushu communities through heritage tourism. In Botswana, and in southern Africa more generally, World Heritage status is being used for economic development; this can be somewhat paradoxical to the conservation aims of the 1972 World Heritage Convention (UNESCO 2005). Here I demonstrate that World Heritage status is a major juncture in the livelihoods of the Ju/'hoan and Hambukushu communities living near the Tsodilo World Heritage Site, a heritage site contested between the two ethnic groups and located in a rural, underdeveloped area of Botswana.

Tsodilo World Heritage Site

In the northwest corner of Botswana, about 40 kilometers west of the Okavango Delta, lie four hills: Male, Female, Child, and Grandchild. Together they are now known as the Tsodilo Hills and are recognized for one of the highest concentrations of rock art in the world, as well as an approximately 100,000-year-long archaeological record of habitation spanning the Middle Stone Age to the present (see Campbell et al. 2010). Tsodilo's archaeological heritage and the intangible heritage associated with it are the reasons why the hills were designated as World Heritage in 2001 (criteria i, iii, and vi) (UNESCO 2001).[2] Taken together, the increase in tourism and two recent heritage management plans associated with Tsodilo's World Heritage status (i.e., Campbell 1994; Ecosurv 2005) have had an enormous impact on the identity politics of the Ju/'hoansi and Hambukushu living beside the hills. Understanding the historical background of tourism and heritage management at Tsodilo is necessary to contextualize the current situation.

Explorers, researchers, and tourists have been coming to Tsodilo for over a hundred years (Passarge 1907; Balsan 1955; Hoare 2007; see also Wilmsen 1997). The most famous explorer to come to Tsodilo was Sir Laurens van der Post, a charismatic South African author (and spiritual advisor to Prince Charles) who deeply romanticized 'Bushman' culture (see Jones 2001).[3] Van der Post led an expedition team to Tsodilo in 1955, and his subsequent BBC television series (1956) and book (1958) based on this expedition – both extremely popular – helped to root the place in Western consciousness. Van der Post's stories, along with travel accounts of Tsodilo produced by other explorers, undoubtedly inspired a wave of researchers, including professional and amateur archaeologists and anthropologists who, since the 1960s, have researched and published on the rock art and archaeology of Tsodilo and the oral history related to the site (see Campbell et al. 1994a; Campbell et al. 1994b; Campbell et al. 2010). Researchers, in turn, through their publications, gave the site's archaeology and inhabitants more

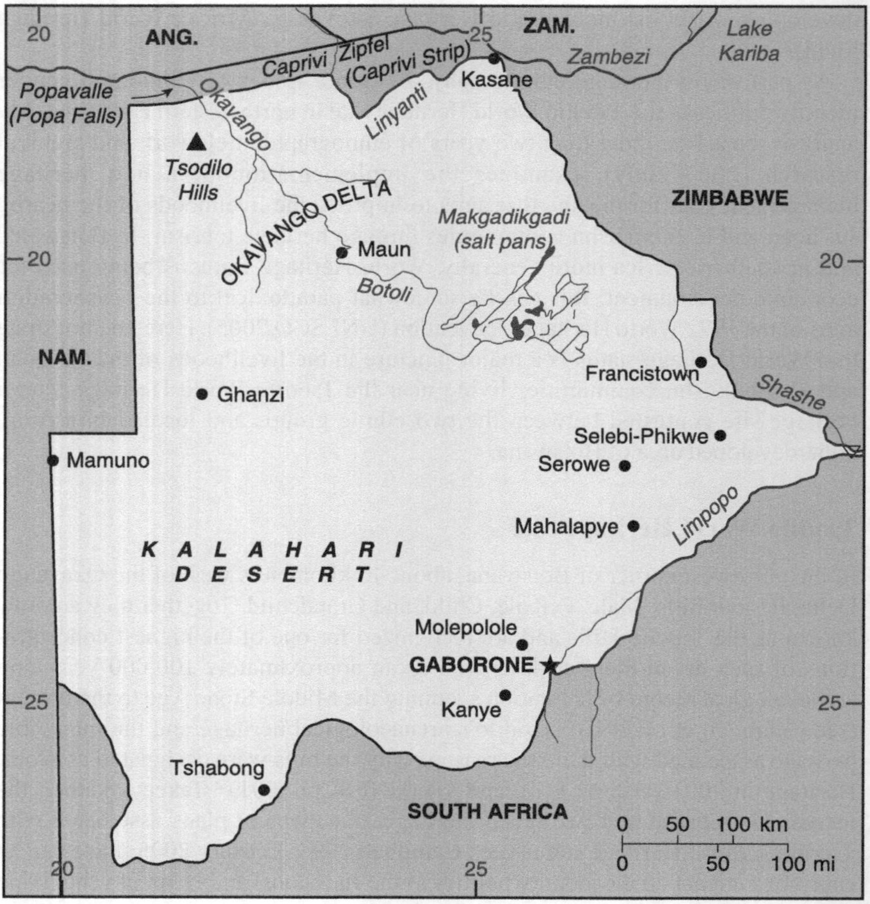

Figure 8.1 The Tsodilo World Heritage Site is in the far northwest of Botswana.

Source: *CIA World Factbook* (2010).

credibility and helped to transform Tsodilo into a tourist destination. Indeed, the draw of Tsodilo for researchers and tourists alike is the opportunity to gaze upon the cultural and temporal 'Other.'

The media hype from explorer and researcher narratives of Tsodilo opened up a new market for Tsodilo: tourism. This was an opportunity not lost on the Ju/'hoansi and Hambukushu, the two ethnic groups living at Tsodilo in the 1960s. Starting in 1969, tourism operators began coming to Tsodilo with mostly wealthy foreign tourists who wanted to see the newly famed rock art as well as meet and buy crafts from the elusive 'Bushmen,' represented at the site by the Ju/'hoansi.[4] About this time, archaeologists already began to formally extrapolate theories of the link between southern African rock art and the ancestors of present-day

Khoisan speakers (see Rudner 1965; Rudner and Rudner 1970), and Tsodilo became one of the few places left where 'Bushmen' could be found alongside rock art.[5] Elsewhere in southern Africa, Khoisan speakers were annihilated, assimilated, or displaced by whites and Bantu speakers.

In the 1970s, the nomadic Ju/'hoansi settled more permanently at the base of Male Hill (Taylor 1997). The Hambukushu also increased their numbers at Tsodilo by encouraging family members to relocate there from villages along the Okavango Delta. The growth of Tsodilo's settlement is, in part, due to the economic benefits of tourism. The hills were so remote and difficult to get to that tourists usually had to hire qualified safari guides to take them there by small-engine planes or with well-equipped four-wheel drive vehicles, which from the main road took at least three or four hours (and typically caused engine overheating and flat tires). Nonetheless, throughout the 1970s, 1980s, and 1990s, the very small stream of tourists to Tsodilo steadily increased. In 1988 tourist visitor numbers to Tsodilo were about 1,000 and in 1996 about 2,000 (Walker 1998: 5). The number of domestic and regional tourists also increased; they primarily came for educational visits (i.e., school groups) and religious purposes (i.e., church groups), such as for prayer and to collect sacred water from the natural springs found in the hills.[6] Tour guiding and the production and sale of tourist crafts became a significant source of income for the Ju/'hoansi, but tourism was, at the time, a less significant source of income for the Hambukushu as only a few were involved in tour guiding.[7]

Relations between the Ju/'hoansi and the Hambukushu became even more strained when tourism became a central contributor to the Tsodilo economy. The Ju/'hoan household economy became more dependent on tourism profits once they settled permanently at Male Hill because they lost the range of their hunting and foraging areas due to the growing number of settlers from other areas of Botswana encroaching on their traditional territories and because of new land policies ratified in the 1970s (see Hitchcock et al. 2009).[8] The Ju/'hoansi also lost their hunting rights when the government of Botswana enacted the requirement for special hunting licenses that were hard to get (see Hitchcock 2001), and as of 2014 hunting was entirely banned. However, the Hambukushu household economy was primarily based on horticulture and herding, and tourism profits were more peripheral. Due to their historical allegiance to the ruling Batawana (Tlou 1985) – a tribe of the Tswana who form the ethnic majority and elites of Botswana – the Hambukushu headman was made chief of the entire Tsodilo area (granting him legal authority over the Ju/'hoansi). This led to the attempted domination of the Ju/'hoansi by the Hambukushu at Tsodilo, later exacerbated by development initiatives such as community-based heritage tourism. For example, in the past the Hambukushu took tourism profits from the Ju/'hoansi, claiming that they were their 'agents.' They also sometimes refused the Ju/'hoansi water from a well near the Hambukushu settlement and occasionally had physical altercations with them. The subjugation of the Ju/'hoansi was part of a larger, more systematic marginalization of Khoisan speakers in Botswana (see Hitchcock and Holm 1993).

Starting in the late 1980s, archaeologists and administrators from the Botswana National Museum pressed for international recognition to conserve Tsodilo, specifically to achieve UNESCO World Heritage status for the site. Part of the process of developing Tsodilo for heritage conservation, as outlined in the 1994 heritage management plan, involved fencing the hills and relocating the Ju/'hoan population outside the new conservation zone. The Hambukushu also lost cattle-grazing lands near the base of Male Hill. In 1994 the Ju/'hoansi were relocated several kilometers away from the hills, which were not only their primary source of income but also the area in which they foraged for food and had nearly year-round water from the natural springs. The Ju/'hoansi were relocated from the lush flora and more abundant fauna of the hills to exposed scrub savannah where local food and medicinal plant resources were not as readily available. The government drilled a water borehole at the new location to entice the Ju/'hoansi, but over time the Hambukushu squabbled with the Ju/'hoansi, claiming control of it. Now the Ju/'hoansi are further from Tsodilo than the Hambukushu, though they are more dependent on tourism and they are the ethnic group most tourists want to see.

Botswana finally signed the 1972 World Heritage Convention in 1997, and Tsodilo was the first heritage site that the new State Party wanted listed (see National Museum, Monuments and Art Gallery 2000). The World Heritage Centre requested that Botswana prepare an updated heritage management plan for

Figure 8.2 The gate to the Tsodilo World Heritage Site with Male Hill (right) and Female Hill (left) in the background. Photo by author, 2005.

Tsodilo, one to be devised in consultation with local community members. In an attempt to secure funding for a new heritage management plan, the Botswana National Museum collaborated with an NGO umbrella, the Kuru Family of Organizations (KFO).[9] KFO secured funding from De Beers, a large diamond corporation, but in return for their assistance in obtaining the funding KFO required that the new heritage management plan offer opportunities for sustainable development to the local community members (see Trust for Okavango Cultural and Development Initiatives 2002a, 2002b).[10] A second heritage management plan was finally completed in July 2005 (Ecosurv 2005).

The 2005 heritage management plan for Tsodilo is expected to improve the livelihoods of the approximately 200 Ju/'hoansi and Hambukushu living there through community-based heritage tourism.[11] In 2001 the site received 2,301 visitors and in 2005 it received 10,362 (Segadika 2010: 146). Numbers have continued to rise to more than 12,000 visitors a year. This exponential increase is undoubtedly due to better accessibility to the site.[12] Now that Tsodilo is becoming a more established tourist destination, plans are underway for private tourism companies to build lodges near the hills (as outlined in the 2005 heritage management plan). A community-run campground was erected in 2013 to help generate more revenue for local community members.

Heritage tourism and ethnic identity politics

Ethnicity is increasingly becoming a global commodity and is also used for such purposes as branding (e.g., Cajun cuisine) and corporate organization (e.g., Native American casinos). On this subject, John L. and Jean Comaroff (2009: 24) write that 'ethno-commerce feeds an ever more ubiquitous mode of production and reproduction, one born of a time in which . . . the sale of culture has replaced the sale of labor in many places.' One of the most recognizable ways in which ethnicity becomes a product or service is through heritage tourism, where an ethnic group's otherness is desirable and consumed. However, to keep their appeal to tourists, members of an ethnic group profiting from heritage tourism need to maintain (or produce) their cultural difference. Thus, local communities living near World Heritage Sites that are designated with criterion vi for intangible heritage – reflecting the beliefs and oral histories of these communities – also need to conserve their intangible heritage (or at least the appearance of an authentic intangible heritage) to ensure a more long-term engagement in heritage tourism.

Exoticized representations of ethnic groups entice tourists to seek out foreign cultures through travel. Indigenous ethnic groups are of special interest because they are considered the ultimate 'Other' and because their traditional lifestyle is endangered. For example, the Ju/'hoansi are romantically portrayed as primitive 'Bushmen' in popular culture and, like other Khoisan speakers, are considered indigenous to southern Africa.[13] The Ju/'hoansi at Tsodilo are aware of tourists' fascination with them as 'Bushmen' and indigenes. They are able to profit from this perception by attracting more tourists, who, in addition to touring the hills, want to take photos of the Ju/'hoansi and buy crafts from them. This seemingly puts the Ju/'hoansi at an advantage over the Hambukushu, who are not always differentiated from other Bantu speakers, when it comes to staking a claim in the heritage tourism market. Furthermore, their presence at a rock art site – the imagery attributed to their ancestors – is unique in southern Africa and is thus itself a huge touristic draw for Tsodilo.

Appearances are deceiving, though, as the Ju/'hoansi and Hambukushu are both relative newcomers to Tsodilo. The Ju/'hoansi and Hambukushu have at least occasionally inhabited Tsodilo since the late nineteenth century. The Ju/'hoansi are traditionally nomadic, pursuing the game, vegetation, and water needed to survive in the harsh Kalahari Desert. When they arrived at Tsodilo in the late nineteenth century following the invasion of the Batawana in the area south of Tsodilo, they met the Khwe (also Khoisan speakers) who were eventually displaced to the Okavango Delta. The Hambukushu came to Tsodilo at about the same time, fleeing from the north where their chiefs sold them into the transatlantic slave trade (Larson 2001; Tlou 1985). At first, they depended on subsistence foraging, but by the mid-twentieth century they had settled along the western edges of the Okavango Delta, where they established horticultural areas, fished, and herded.

Heritage tourism is an interesting phenomenon in Botswana, a culturally hegemonic country. The government does not officially recognize non-Tswana tribes, or even ethnic difference, and this stance benefits the Tswana elites through the denial of inequality (see Nyati-Ramahobo 2008).[14] However, tourism is now the second biggest contributor to the country's gross domestic product in the country behind mining (World Travel and Tourism Council 2007), and heritage tourism, an industry that markets ethnic difference, is also becoming a significant contributor to the national economy. Botswana's ethnic minorities, especially Khoisan speakers, potentially have the opportunity for economic gain and improved political representation through the expansion of the heritage tourism sector.

The Ju/'hoansi and Hambukushu at Tsodilo, both relatively impoverished by the country's standards, have a fraught relationship as they compete for resources. In the past, they competed over access to water at the hills. Over time, this included ownership of the hills and control over the economic benefits of heritage tourism. However, as the Hambukushu are the more dominant ethnic group at Tsodilo, they control most of the development initiatives instigated by the government of Botswana and KFO, including some of the projects outlined in the 2005 heritage management plan.[15] Community-based heritage tourism actually exacerbates ethnic tensions as the Ju/'hoansi and Hambukushu are encouraged to commodify their identities to attract the interest of tourists in order to profit from this market. This is also evident in the ways that the Ju/'hoansi and Hambukushu recall their origin stories and in the interpretations of Tsodilo's rock art that Ju/'hoan and Hambukushu tour guides relay to tourists.

The Ju/'hoansi and Hambukushu interpret Tsodilo's rock imagery differently as they guide tourists. During my fieldwork, I observed that the Ju/'hoan guides describing the rock art for tourists interpret the images that researchers refer to as

Figure 8.3 Rock imagery at Tsodilo generally referred to as 'geometric designs' by scholars is interpreted and relayed to tourists by Ju/'hoan guides as 'tortoise shells' and 'suns' and by Hambukushu guides as 'baskets.' Photo by author, 2005.

'geometric designs,' as 'tortoise shells' or 'suns.' However, the Hambukushu guides interpret these same rock art images for tourists as 'baskets,' which Hambukushu women make. Meanwhile, rock art researchers seek to emphasize their scientific interpretations through educating the local tour guides about their research findings on the rock art (e.g., the Botswana National Museum hosts workshops for local tour guides and site museum staff). This includes attributing the imagery (and other archaeological material) to the ancestors of both Khoisan speakers and Bantu speakers, an interpretation that the government encourages in an attempt to nationalize Tsodilo's cultural heritage by not privileging Khoisan speaker origins.[16] Thus researchers are influencing local tour guides to use their more scientific understanding of the rock art, which undermines local interpretations of rock art that are part of Tsodilo's intangible heritage.

In addition, the Ju/'hoansi and Hambukushu both claim Tsodilo as their creation site.[17] They point to the same groove marks that are found on large boulders on the top of Female Hill as evidence of their origins. The Ju/'hoansi claim these groove marks to be the tracks of the first animals on earth when the hills were 'new' and therefore more malleable. Some of the groove marks are said to have been made by the eland, a large African antelope that is important in Ju/'hoan (and other Khoisan) mythology (see Lewis-Williams and Pearce 2004). The Hambukushu, however, consider these groove marks to be the tracks of the cow, which is important in their cultural worldview (see Segadika 2006: 34). In fact, Hambukushu from neighboring countries also believe Tsodilo to be their origin site, and Hambukushu tribal leaders from Namibia and Angola visit the hills to pay respect. Increased tourism and the benefits that can be reaped from the industry, especially now that the site is World Heritage, have both ethnic groups competing for the accuracy of their origin stories. These origin stories, which are part of their intangible heritage and a reason for the use of criterion vi in the site's World Heritage bid, are now increasingly commodified by the Ju/'hoansi and Hambukushu, as well as tour operators, to attract tourists.

Tourism development and gender

Tourism development is also a catalyst for changing gender roles among traditional communities conscripted into a market economy. At Tsodilo, the most obvious factors influencing the gender roles of the Ju/'hoansi and Hambukushu are assimilation into a national, Tswana culture and their integration into a cash economy. The HIV/AIDS pandemic is also taking its toll on the Ju/'hoansi and Hambukushu, affecting the social structure of their families. Indeed, Botswana has the second highest rate of HIV/AIDS in the world (UNAIDS 2009). Although there are a number of noticeable disruptive forces challenging the traditional gender roles of the Ju/'hoansi and Hambukushu, heritage tourism is significant in that it affects both labor division and cultural representations of gender.

With their traditional livelihoods in flux, so are their gender relations. The Ju/'hoansi are traditionally an egalitarian society, and decisions are made together as a group (Lee 1979). The basic division of labor between genders centers on food procurement: Ju/'hoan men hunt large animals and Ju/'hoan women gather plant resources. No longer able to practice this lifestyle, Ju/'hoan men are now lucky if they find employment working at a cattlepost or are chosen for piecemeal jobs offered by the government as a form of rural development. Ju/'hoan women also work these piecemeal jobs, and both men and women collect state welfare. The Hambukushu are traditionally a patriarchal, though matrilineal, society (Larson 2001). Men have more authority and are actively involved in making community decisions. Although the chief is a man, the chiefly line passes down to the son of the current chief's older sister. Hambukushu women do not usually engage in leadership roles beyond the domestic sphere of their household compounds. As horticulturalists, Hambukushu men are responsible for larger crops, such as sorghum and millet, and women are responsible for smaller crops, such as legumes and pumpkins. Men also herd cattle and goats and hunt (though it is also now illegal for them to hunt). Hambukushu women look after the smaller animals of the household (e.g., chickens). Although the Hambukushu are still able to practice their traditional way of life, they also seek upward mobility through employment.

As I learned at Tsodilo, social and economic development via heritage tourism contributes to the transformation of the gender roles of Botswana's rural communities and ethnic minorities in multiple ways. For example, the government of Botswana and KFO approach community development and village representation through the formation of community trusts. These community trusts are based on geography and therefore often represent an ethnically heterogeneous population, so they potentially agitate existing social politics between and within these communities. As gender is intertwined with ethnicity, gender relations are also disrupted. Although Botswana does not recognize ethnic difference, its policies privilege Tswana patriarchal culture much to the detriment of egalitarian Khoisan speakers. The Tsodilo Community Development Trust was registered in December 2005 in order to represent the entire population at Tsodilo in development projects stemming from the 2005 heritage management plan. Even though they are culturally quite distinct and have a history of fraught relations, the Ju/'hoansi and Hambukushu must now work together in making decisions for their communities in order to abide by the principles of the 2005 heritage management plan.

The Tsodilo Community Development Trust is a point of contention between the Ju/'hoansi and Hambukushu. The traditionally egalitarian Ju/'hoansi now face social stratification due to the community trust's hierarchical structure where board members are voted in and receive financial allowances, making representation highly competitive. The Hambukushu, who outnumber the Ju/'hoansi, have dominated the board since the community trust was formed (until KFO more closely monitored the board from mid-2009). Hambukushu men compete to head the community trust's board as it gives them greater social standing and power in controlling the resources for which the community trust is responsible. More Ju/'hoan men participate in the community trust because when Ju/'hoan women participate they are harassed by the Hambukushu men (e.g., accusations of witchcraft). Hambukushu women more regularly participate in the community trust, but Hambukushu men also intimidate them if they seek higher positions on the board. Hambukushu men are increasingly controlling most aspects of Tsodilo's development and heritage tourism industry.

At Tsodilo, the division of labor in the heritage tourism industry reveals that, in general, men become the tour guides while women become the producers of tourist crafts. The majority of Ju/'hoan women produce crafts to sell to tourists (e.g., leather bags and necklaces made from ostrich eggshells and other local materials). A large number of Hambukushu women also produce crafts (e.g., woven baskets and necklaces made from local materials). Many Ju/'hoan men make crafts (e.g., bow and arrow sets, walking sticks, and carved mokolwane palm nuts), but only a few Hambukushu men do so (e.g., wooden spoons and other carved wooden crafts). However, tour guiding is almost completely dominated by men (only one Ju/'hoan woman occasionally guides), especially Hambukushu men. However, men also indirectly control the sale of tourist crafts through their dominance on the community trust board, which manages the craft shop at the site museum. Although most tourists come to Tsodilo to see 'Bushmen' along with rock art, the Hambukushu men are taking over the tour guiding industry in Tsodilo because they are more educated and speak better English. As they transition from their traditional livelihoods to a cash economy through heritage tourism, the Ju/'hoan men are seemingly emasculated (as craft production becomes feminized labor), and women of both ethnic groups are also kept in the 'backstage' of this industry.

Heritage tourism contributes to the changing traditional gender roles among the Ju/'hoansi and Hambukushu by affecting the labor division between men and women as well as the cultural representations of gender for both ethnic groups. Many rural women earn their household income through tourism. Increasingly, women are the heads of households in Botswana due to waning marriage traditions, urbanization, and the HIV/AIDS pandemic. Women are actually the largest demographic group involved in the heritage tourism industry at Tsodilo through the production and sale of tourist crafts, but men control the industry through their leadership in the community trust. This is, in large part, due to the intervention of government and NGO employees. The two ethnic groups represented on the Tsodilo Community Development Trust have very different traditional gender systems, and when forced to work with each other, both traditional cultures are put under strain.

Conclusion

What lessons does heritage tourism at Tsodilo have for scholars and heritage managers working at other World Heritage Sites? World Heritage status is not always perceived of as a conservation strategy, but some countries, as demonstrated by Botswana, see it as a mechanism for economic development. The local communities living near Tsodilo are ethnic minorities and are marginalized by the government of Botswana due, in part, to the country's assimilationist policies. The government appropriates local heritage under nationalism, at the same time promoting cultural commodification in order to encourage more economic responsibility among its rural poor. This use of World Heritage status and the subsequent heritage management plans affect, somewhat problematically, both the ethnic and gendered identity politics of the Ju/'hoansi and Hambukushu living nearby. The use of heritage for development, especially cultural heritage, has important implications for the local communities residing near World Heritage Sites, and as the sociopolitical dynamics of World Heritage Sites are unique they each deserve more critical inquiry. Heritage ethnographies enable a more nuanced understanding of these impacts, for example on the reconfiguration of ethnicity and gender, which are relevant to scholarship on World Heritage and tourism, and, of course, to ongoing heritage management at these sites.

Notes

1 Based on his work on community forests in Cameroon, where he realized that the village becomes the unit of community in the eyes of the law, Peter Geschiere (2009: 77) asks: 'why do . . . developers see it as self-evident that in Africa development has to be realized through "communities"?'

2 Botswana proposed criterion vi for Tsodilo on the grounds that 'traditions speak of Tsodilo as being the home of all living creatures, more particularly home to the spirits of each animal, bird, insect, and plant that has been created. Though exact interpretation and dating of the rock art are uncertain, the art itself clearly testifies to the long tradition of the site as spiritual, a tradition continued today in practices of the !Kung and in visits by, in effect, pilgrims in Western parlance, often from some distance' (!Kung is another name for Ju/'hoansi) (UNESCO 2001: 61). ICOMOS, a World Heritage Advisory Body, evaluated Tsodilo's fulfillment of criterion vi as follows: 'the Tsodilo outcrops have immense symbolic and religious significance for the human communities who continue to survive in this hostile environment' (UNESCO 2001: 65).

3 'Bushman' is a settler concept used to refer to Khoisan speakers. In Botswana, the government and many citizens refer to Khoisan speakers as 'Basarwa' (see Motzafi-Haller 1994). Both of these terms are still in use but are now considered pejorative. In the 1960s, anthropologists adopted the use of the linguistic term 'San' to refer to Khoisan speakers (see Hitchcock and Biesele 2012). I use the term 'Khoisan speakers' to represent all the cultures of the click-consonant language family because in the region where I work, Ngamiland, both Khoe and San languages are spoken. Otherwise, I refer to particular Khoisan ethnic groups by their language name (e.g., Ju/'hoansi and Khwe).

4 The exclusive Lindblad Expeditions operated photographic safaris to Tsodilo from 1969 to the early 1980s. Other private safaris, such as those organized by the Shakawe Fishing Camp, which is still in operation (Shakawe River Lodge), ventured to Tsodilo throughout the 1970s and 1980s. Many more tourism operators now take their clients to Tsodilo.

5 This theory was further expanded upon by the work of J. David Lewis-Williams in his seminal book *Believing and Seeing: Symbolic Meanings in Southern San Rock Paintings* (1981), which was informed, in part, by the work of contemporary ethnographers working with Khoisan speakers.

6 Church groups, such as the Zion Christian Church, visit Tsodilo to pray and collect water from the sacred well 'Chokamo' (also referred to as 'Tshokgam' by Campbell et al. 2010: 27).

7 The Hambukushu chief, who died in 2013, was most likely the guide who led van der Post around Tsodilo during his 1955 expedition. The Hambukushu chief, his son, and one or two older Hambukushu men also guided a number of researchers over the past several decades. Recently, though, several other Hambukushu—all young men—have taken up tour guiding at Tsodilo to earn an income.

8 After independence in 1966, the new government of Botswana created tribal land boards that only recognized Tswana tribal territories. Citizens were encouraged to settle anywhere in the country through application to these land boards. The government especially encouraged applications for cattleposts, as cattle rearing is the traditional livelihood of the pastoral Tswana. However, cattleposts require a lot of land and they quickly degrade the environment through overgrazing.

9 The two KFO NGOs involved in the new heritage management plan for Tsodilo are the Letloa Trust, which is the administrative NGO of KFO, and the Trust for Okavango Cultural and Development Initiatives (TOCaDI), which is a regional NGO in Ngamiland focusing on community development.

10 De Beers worked with Debswana (a joint venture between De Beers and the government of Botswana) to create the Diamond Trust, which is a 'community social investment' organization. The Diamond Trust's first investment recipient was the new heritage management plan for Tsodilo, for the implementation of which the Diamond Trust pledged 10 million Botswana Pula (about 1.5 million US dollars).

11 The project has slowly been implemented since the heritage management plan was finalized in July 2005, though the money was only released to KFO in November 2009.

12 One of the sandy paths leading to Tsodilo from the main asphalt road was graveled by early 2001 due to the official opening of the site museum and in anticipation of World

Heritage status. This made it possible to drive to the new site museum from the main road in less than an hour and with two-wheel drive vehicles.

13 In addition to academic scholarship, the Ju/'hoansi also feature in photography books (e.g., Johnson et al. 1979), popular documentary films produced by John Marshall, and notably in the movie series *The Gods Must Be Crazy* (Uys 1980).

14 For example, the application of the term 'indigenous' is contested in Botswana. The government considers all of its citizens 'indigenous,' though most international and regional human rights groups reserve this status for Khoisan speakers and their descendants (e.g., the International Work Group for Indigenous Affairs, the Indigenous Peoples of Africa Co-ordinating Committee, and the Working Group of Indigenous Minorities in Southern Africa).

15 For example, the government has a Village Development Committee project in Tsodilo, and the board is almost entirely composed of Hambukushu. The site museum at Tsodilo employs many more Hambukushu than Ju/'hoansi. Also, a craft shop at the site museum, which TOCaDI helped to facilitate and originally employed one Hambukushu woman and one Ju/'hoan woman, was over the course of two years (2007–9) taken over by Hambukushu who excluded the Ju/'hoansi from further participation.

16 Tsodilo is a pivotal site in the contentious Kalahari Debate, which polarizes anthropologists and archaeologists over how much interaction Khoisan speakers had with Bantu speakers prior to recent history (for an overview see Sadr 1997), and the debate has had modern-day political consequences. Some archaeologists and anthropologists use interpretations of Tsodilo's past to argue that Bantu speakers have long interacted (and intermarried) with Khoisan speakers (see Wilmsen 1989).

17 The displaced Khwe also consider Tsodilo their creation site (Chumbo and Mmaba 2002: 8).

References

Balsan, F. (1955) *Capricorn Road*, trans. P. Search, New York: Philosophical Library.

Breglia, L.C. (2006) *Monumental Ambivalence: The Politics of Heritage*, Austin: University of Texas Press.

Campbell, A. (1994) *Tsodilo Hills Management Plan: Scheme for Implementation*, Gaborone: Government Printer.

Campbell, A., Denbow, J. and Wilmsen, E. (1994a) 'Paintings Like Engravings: Rock Art at Tsodilo', in T.A. Dowson and D. Lewis-Williams (eds.) *Contested Images: Diversity in Southern African Rock Art Research*, Johannesburg: Witwatersrand University Press.

Campbell, A., Robbins, L.H. and Murphy, M.L. (1994b) 'Oral Traditions and Archaeology of the Tsodilo Hills Male Hill Cave', *Botswana Notes and Records*, 26: 37–54.

Campbell, A., Robbins, L. and Taylor, M. (eds.) (2010) *Tsodilo: Copper Bracelet of the Kalahari*, East Lansing: Michigan State University Press.

Chumbo, S. and Mmaba, K. (eds.) (2002) *||Xom Kyakyare Khwe: ǂAm Kuri Kx'ûî â* (The Khwe of the Okavango Panhandle: The Past Life), Shakawe: The Teemacane Trust.

Cleere, H. (2001) 'The Uneasy Bedfellows: Universality and Cultural Heritage', in R. Layton, P.G. Stone and J. Thomas (eds.) *Destruction and Conservation of Cultural Property*, One World Archaeology, 41, London: Routledge.

Comaroff, J.L. and Comaroff, J. (2009) *Ethnicity, Inc.*, Chicago, IL: University of Chicago Press.

de Jong, F. and Rowlands, M. (eds.) (2007) *Reclaiming Heritage: Alternative Imaginaries of Memory in West Africa*, Walnut Creek, CA: Left Coast Press Inc.

Di Giovine, M.A. (2008) *The Heritage-scape: UNESCO, World Heritage, and Tourism*, Lanham, MD: Lexington Books.

Ecosurv (2005) *Tsodilo Hills World Heritage Site Integrated Management Plan*, Gaborone: Ecosurv.

Fabricius, C., Koch, E., Magome, H. and Turner, S. (eds.) (2004) *Rights, Resources and Rural Development: Community-Based Natural Resource Management in Southern Africa*, London: Earthscan.

Fontein, J. (2006) *The Silence of Great Zimbabwe: Contested Landscapes and the Power of Heritage*, Walnut Creek, CA: Left Coast Press Inc.

Geschiere, P. (2009) *The Perils of Belonging: Autochthony, Citizenship and Exclusion in Africa and Europe*, Chicago, IL: University of Chicago Press.

Harrison, D. and Hitchcock, M. (eds.) (2005) *The Politics of World Heritage: Negotiating Tourism and Conservation*, Clevedon, UK: Channel View Publications.

Hitchcock, R.K. (2001) '"Hunting Is Our Heritage": The Struggle for Hunting and Gathering Rights among the San of Southern Africa', in D.G. Anderson and K. Ikeya (eds.) *Parks, Property and Power: Managing Hunting Practice and Identity within State Policy Regimes*, Senri Ethnological Studies, 59, Osaka: National Museum of Ethnology.

Hitchcock, R.K. and Biesele, M. (2012) San, Khwe, Basarwa, or Bushmen? Terminology, Identity, and Empowerment in Southern Africa. Online, www.kalaharipeoples.org/downloads/San, Khwe, Basarwa, or Bushman.doc (accessed 3 January 2012).

Hitchcock, R.K. and Holm, J.D. (1993) 'Bureaucratic Domination of African Hunter-Gatherer Societies: A Study of the San in Botswana', *Development and Change*, 24(1): 1–35.

Hitchcock, R.K., Biesele, M. and Babchuk, W. (2009) 'Environmental Anthropology in the Kalahari: Development, Resettlement, and Ecological Change among the San of Southern Africa', *Explorations in Anthropology*, 9(2): 180–188.

Hoare, M. (2007) *Mokoro: A Cry for Help!*, Durban: Partners in Publishing.

Johnson, P., Bannister, A. and Wannenburgh, A. (1979; 2nd ed. 1999) *The Bushmen*, Cape Town: Struik Publishers.

Jones, J.D.F. (2001) *Storyteller: The Many Lives of Laurens van der Post*, London: John Murray.

Kirshenblatt-Gimblett, B. (2006) 'World Heritage and Cultural Economics', in I. Karp, C.A. Kratz, L. Szwaja and T. Ybarra-Frausto, with G. Buntinx, B. Kirshenblatt-Gimblett and C. Rassool (eds.) *Museum Frictions: Public Cultures/Global Transformations*, Durham, NC: Duke University Press.

Larson, T.J. (2001) *The Hambukushu Rainmakers of the Okavango*, Lincoln, NE: Writers Club Press.

Leask, A. and Fyall, A. (eds.) (2006) *Managing World Heritage Sites*, Oxford: Elsevier.

Lee, R.B. (1979) *The !Kung San: Men, Women, and Work in a Foraging Society*, Cambridge: Cambridge University Press.

Lewis-Williams, J.D. (1981) *Believing and Seeing: Symbolic Meanings in Southern San Rock Paintings*, London: Academic Press.

Lewis-Williams, J.D. and Pearce, D.G. (2004) *San Spirituality: Roots, Expressions, and Social Consequences*, Walnut Creek, CA: AltaMira Press.

Motzafi-Haller, P. (1994) 'When Bushmen Are Known as Basarwa: Gender, Ethnicity, and Differentiation in Rural Botswana', *American Ethnologist*, 21(3): 539–563.

National Museum, Monuments and Art Gallery (2000) *Tsodilo, Mountain of the Gods: World Heritage Nomination Dossier*, Gaborone: National Museum, Monuments and Art Gallery.

Nyati-Ramahobo, L. (2008) *Minority Tribes in Botswana: The Politics of Representation*, London: Minority Rights Group International.

Owens, B.M. (2002) 'Monumentality, Identity, and the State: Local Practice, World Heritage, and Heterotopia at Swayambhu, Nepal', *Anthropological Quarterly*, 75(2): 269–316.

Passarge, S. (1907) *Die Buschmänner der Kalahari*, Berlin: Dietrich Reimer (Ernst Vohsen).

Pedersen, A. (2002) *Managing Tourism at World Heritage Sites: A Practical Manual for World Heritage Site Managers*, Paris: World Heritage Centre.

Reed, M.G. (1997) 'Power Relations and Community-Based Tourism Planning', *Annals of Tourism Research*, 24(3): 566–591.

Rudner, I. (1965) 'Archaeological Report on the Tsodilo Hills, Bechuanaland', *South African Archaeological Bulletin*, 20(78): 51–70.

Rudner, J. and Rudner, I. (1970) *The Hunter and His Art in Southern Africa*, Cape Town: C. Struik.

Sadr, K. (1997) 'Kalahari Archaeology and the Bushman Debate', *Current Anthropology*, 38(1): 104–112.

Schroeder, R.A. (1999) *Shady Practices: Agroforestry and Gender Politics in the Gambia*, Berkeley: University of California Press.

Segadika, P. (2006) 'Managing Intangible Heritage at Tsodilo', *Museum International*, 58(1–2): 31–40.

—— (2010) 'The Power of Intangible Heritage: Bottom-Up Management', in A. Campbell, L. Robbins and M. Taylor (eds.) *Tsodilo Hills: Copper Bracelet of the Kalahari*, East Lansing: Michigan State University Press.

Smith, L. (2006) *Uses of Heritage*, London: Routledge.

Taylor, M. (1997) 'These Are Our Hills: Processes of History, Ethnicity, and Identity among Ju/'hoansi at Tsodilo', Paper presented at the Khoisan Identities and Cultural Heritage Conference, July 12–16, 1997.

Titchen, S.M. (1996) 'On the Construction of "Outstanding Universal Value": Some Comments on the Implementation of the 1972 UNESCO World Heritage Convention', *Conservation and Management of Archaeological Sites*, 1(4): 235–242.

Tlou, T. (1985) *A History of Ngamiland 1750 to 1906: The Formation of an African State*, Gaborone: Macmillan Botswana Publishing Company (Pty.) Ltd.

Trust for Okavango Cultural and Development Initiatives (2002a) *Tsodilo Community Initiative I: Community Projects for the People of Tsodilo*, Shakawe: Trust for Okavango Cultural and Development Initiatives.

—— (2002b) *Tsodilo Community Initiative II: A Suggestion to Include a Comprehensive EIA in the New Management Plan for Tsodilo Hills, Including an Evaluation of the Herein Presented Concept Community Development Alongside Flora/Fauna Conservation*, Shakawe: Trust for Okavango Cultural and Development Initiatives.

Tucker, H. (2003) *Living with Tourism: Negotiating Identities in a Turkish Village*, London: Routledge.

UNAIDS (2009) *AIDS Epidemic Update: November 2009*, Geneva: UNAIDS.

UNESCO (2001) *Tsodilo (Botswana) No. 1021 (Advisory Body Evaluation)*, Paris: UNESCO.

—— (2005) *Basic Texts of the 1972 World Heritage Convention, Edition 2005*, Paris: UNESCO.

Uys, J. (1980) *The Gods Must Be Crazy*, movie series, Johannesburg: Ster-Kinekor Pictures.

van der Post, L. (1956) *The Lost World of the Kalahari*, TV series, London: BBC.

—— (1958) *The Lost World of the Kalahari*, New York: William Morrow and Company.

Walker, N. (1998) 'Monuments and Tourism: Developing Public Heritage Sites in Botswana', *Zebra's Voice*, 25(1): 4–6.
Wilmsen, E.N. (1989) *Land Filled with Flies: A Political Economy of the Kalahari*, Chicago, IL: University of Chicago Press.
—— (ed.) (1997) *The Kalahari Ethnographies (1896–1898) of Siegfried Passarge: Nineteenth Century Khoisan- and Bantu-Speaking Peoples*, Cologne: Rüdiger Köppe Verlag.
World Travel and Tourism Council (2007) *Botswana: The Impact of Travel and Tourism on Jobs and the Economy*, London: World Travel and Tourism Council.

9 Tourism community involvement strategy for the Living World Heritage Site of Hampi, India

A case study[1]

Bernhard Bauer, Nitin Sinha,
Michele Trimarchi and Vincenzo Zappino

Today tourists' understanding of becoming an active agent in their travel experiences, moving between and among multiple localities, in an environment of transnational, interconnected social networks, is rising. Tourists are perceived as the mobile modern population of the world, soaking up information and life experiences. However, in the huge field of tourism, especially when we come to cultural or eco tourism, we should not only consider the tourists but also the social aspect of interaction with the local population and their environment. The field of sustainable tourism is based on various aspects mainly focusing on environmental and economic issues but also on social and cultural constraints. Strategies about the active involvement of local communities in tourism structures may cover many of the mentioned fields at the same time, if implemented properly.

This study presents a strategy for sustainable community involvement within a general tourism development project at the World Heritage Site of Hampi, India. Sustaining the community has become an essential element of sustainable tourism. Taylor (1995: 487) argues that 'the concept of community involvement in tourism development has moved nearer to the center of the sustainability debate', which means in other words that, without community sustainability, tourism development cannot be expected to be sustainable (Richards 2000). Murphy (1985) stresses that the 'good will and cooperation of host communities' are among the essential elements of community development. The complexity of communities and the relationships between them pose significant challenges for sustainable development of tourism. The strategy puts community individuals at the centre of the discussion concerning the preservation of intangible cultural heritage and the commodities of tangible monuments.

The concept of living cultural heritage (LCH), which is an important asset for the participatory development approach representing an empowerment strategy for the local population, is the theoretical contextualization. The concept is primarily based on three elements for the successful involvement of locals. The *local consultative committees* enhance the co-operation between different actors in the tourism development process; the involvement of locals in the *conservation and preservation process* gives a sense of ownership; and integration into the *interpretation strategy* is a way to promote cultural commitment and pride, capacity building and education.

The United Nations Educational, Scientific and Cultural Organization (UNESCO) World Heritage Site (WHS) of Hampi was put on the World Heritage List in 1986 as a group of monuments. After the completion of a general *Integrated Management Plan* (IMP), a *Master Plan* and the establishment of the Hampi World Heritage Area Management Authority[2] (HWHAMA), the site of Hampi was delisted from the World Heritage in Danger list where it appeared from 1999 until 2006. Hampi is a unique site due to its dimensions, in terms of geographical scale and the immense number of monuments (recently defined as 1,600) and cultural resources.

The area of Hampi is divided into core, buffer and peripheral zones; the core zone is the most populated with almost 47 per cent of the total population (75,000).[3] Hampi Village[4] has registered the highest growth in population in the area due to radical socio-economic transformation and its function as a reception centre for foreign tourists. While the greater part of the active population has traditionally been involved in the agricultural production of sugar cane, rice, banana and jiggery, in recent years tourism has become an increasingly important sector, revitalizing the local economy and attracting new investments especially in the private sector.

The Tourism Development Strategy for Hampi World Heritage Area (HWHA) was planned by tourism development consultants for the Karnataka Department of Tourism (DoT) and the HWHAMA in 2009–2010. It includes a strategy for general infrastructure, traffic and transportation, legal issues, a marketing-tourism segment and product strategy, environmental protection and community involvement. The approach proposed by the strategy puts local communities and related environmental issues at the centre of the tourism development, which has to become an opportunity for local communities in terms of job creation, additional revenues, heritage preservation and socio-economic development. The strategy is based on an integrated tourist model which promotes the integration of both tourists into the local culture and local inhabitants into the benefits generated by tourism development.

Definition of living cultural heritage and the LCH concept

The discussion about preserving intangible cultural heritage started seriously in 1989 with the *Recommendation on the Safeguarding of Traditional Culture and Folklore* by UNESCO. The prevailing ideas of monumentality and authenticity were followed by the *Proclamation of Masterpieces of the Oral and Intangible Heritage of Humanity* in 1997, which finally resulted in the Convention for the Safeguarding of the Intangible Cultural Heritage (ICH) in 2003 in Paris.

The convention's definition of ICH creates confusion through the synonymous use of the term living heritage.[5] According to Munjeri (1995), the concept of living cultural heritage (LCH) acknowledges that objects and places acquire meaning and significance through the values that are ascribed to them by the people coming into contact with these objects. This change in understanding of cultural heritage could signify a new way of understanding objects and spaces by focusing on the human element associated with them. The convention did not specify living heritage related to monuments and buildings but only to persons *living* their traditions. This concept was additionally strengthened by the Living Human Treasures.[6] According to Keiko Miura's position (2005), living heritage brings opportunities and creates relations between cultural properties and communities in achieving their common future.

While ICH represents traditions, customs, practices and knowledge recognized by communities and groups, the term LCH signifies the linkage of these intangible assets with monuments and sacred places that are actively used for rituals and other forms of (daily) social interaction. The synergy of intangible and tangible cultural heritage merges into living cultural heritage.

The living cultural heritage concept is primarily based on active and long-lasting integration of Hampi's local population in the field of tourism. It aims at raising awareness of cultural heritage among locals and encouraging the society to be an active part of their living heritage, and it is an empowerment strategy to higher social capital. The LCH concept is deeply connected with an educational aspect represented by the cultural heritage awareness campaign focusing on the cultural, historical and social meaning of the locals' cohabitation with monuments. This form of awareness raising is especially important for the sub-strategies of site interpretation and conservation within the community involvement strategy. The concept enhances co-operation between actors in the tourism development process through the establishment of local consultative committees as a discussion platform for every element involved in the local tourism industry. Furthermore, the principle of ownership and active integration in the development process gives a sense of responsibility about Hampi's future to its population and tourism stakeholders.

As the local population is still deeply dependent on agricultural activities,[7] this concept also fosters the close relationship between cultural and natural heritage. The social construction of gender work division, as well as the mutual work of farming and harvesting the fields, are part of the local cultural heritage.

Tourists are also an important element of the LCH concept as their proximity to locals is a fact. The level of authenticity is already very high at Hampi and will rise throughout the years if locals can continue their traditional lifestyle. In recent years international tourists are becoming more interested in foreign cultures than in visiting sights only. This fact strengthens the current development plan for Hampi as it is future oriented and balanced with the present trends in cultural tourism.

Living cultural heritage at Hampi

At Hampi, interaction between the community and monuments is strong since its residents still use ancient culturally significant edifices (now called World Heritage Sites). Temples and shrines that serve for pilgrimage and religious activities are in daily use by Indian tourists as well as for local prayers. The institutions and buildings with the greatest importance for social and cultural life and traditions are located in Hampi Bazaar and Anegundi. Hampi Bazaar is an ancient street which was once part of a thriving temple street market. In recent years the pavilions have been converted into shops, restaurants and hostels. The backs of the pavilions are used by villagers as their homes. Merchants, vendors, locals and tourists enjoy Hampi Bazaar as a centre for social interaction, traditional community life, exchanging information and accommodation facilities. The traditional houses and pavilions are generally functional but the individual attempts at restoration and conservation are often too invasive. The surrounding World Heritage landscape is rich in soil resource and quarries and is the basis for agriculture. Cultural and natural heritage are inseparably linked in day to day activities by the local population, domestic visitors and international tourists. The example of Hampi demonstrates that living cultural heritage refers not only to intangible aspects but also to traditions and rituals that are individually connected to specific buildings and sacred places. The tangible part of the cultural heritage is set into active usage through (intangible) customs, traditions and rituals. As long as the community is traditionally attached to the built heritage in such a way, this form of heritage should be defined as living cultural heritage. The daily usage of buildings should be even more encouraged, and the message for conservation should be *how to do it* instead of *don't do it*. If all the buildings and monuments are no longer used, the danger of *museumification* would arise quickly.

Significance of tourism community involvement at Hampi

Due to the structural changes in the tourism management of HWHA planned for the next decade, community involvement and participation are inevitable for the local population at this stage of the strategy implementation. Firstly, active involvement is a form of future oriented socio-economic and cultural development. Secondly, the site of Hampi will keep its authentic atmosphere, which will positively influence the tourism experience.

Active involvement is a crucial issue in safeguarding the intangible heritage and the traditional architecture, both firmly linked to the community's traditions (concerning conservation). The strategy for participation in conservation and preservation is a decisive method for maintaining social equilibrium.[8] Intangible elements of cultural heritage expressed as community life and values or traditions are an important aspect at Hampi. Local cultural values are best preserved if tourism development is led by the resident population itself. If these values can be integrated into the tourism strategy, locals could adjust their cultural activities to the tourism offer and their actions to the tourism development process.

Furthermore, local communities are highly attracted by the increased income opportunities from tourism activities. The economic incentive is doubtless the most important factor in encouraging families to actively participate in the tourism business. Local investment is important for accumulating local ownership and has to be encouraged because income generated by tourism activities stays almost totally in the local economy.

Concerning the question of the tourist, community involvement is a strategy for bringing tourists closer to locals. The demand for cultural tourism, where travelers are looking for local traditions and culturally significant experiences, is constantly growing (WTO 2004). Direct contact with locals, whether during a paid service or through the atmosphere in the village, gives an authentic feeling. People living at Hampi are respected in a serious way and the area develops a sustainable authentic profile where public life has not been damaged by the tourism industry.

Local consultative committees

The local consultative committees are composed of groups of local representatives, regional governmental institutions, public and private tourism stakeholders, the HWHAMA, site managers and tourism experts, as well as NGOs and other private institutions directly connected to tourism activities and the development strategy at Hampi. The implementation of these committees shall ensure communication and interaction between the authorities and local communities. As recommended by the *Integrated Management Plan for Hampi*, these consultative committees will become a platform for exchanging information where local communities will not just get informed by but also inform the authorities (Thakur 2008).

In order to avoid a fragmented approach, all the critical issues in terms of planning, development, heritage, conservation and tourism development are discussed together. The requirements for education, training and consulting to enter the tourism business world will be clarified to make access easier for locals. The committees represent a specific tool for collecting requests and/or information by members to be submitted to the Department of Tourism, local governments, the private sector or any other local, regional and national stakeholders. This form of 'discussion round table' integrates locals in the mutual decision-making process of tourism development in their direct neighborhood and facilitates access to government stakeholders.

The consultative committees are not only a practical tool to make people aware of their participation in the tourism sector. Their main purpose is to create open discussions where HWHAMA, the local population and tourism experts define problems, solution, needs and other issues related to tourism development on a local level.

Community involvement in conservation and protection works

Hampi is a unique area containing a wide variety of buildings representing ancient architectural styles and constructed using local materials. In particular, the present scenario shows that the urban situation is vulnerable and the residents are not aware of the importance of protection and restoration. Often restorations are carried out individually with materials and architectural forms differing from the traditional ones. As already stressed in the *Master Plan*[9] and the Joint Conservation Programme, a correct development strategy for new buildings and services has to be implemented. The fragile structure of Hampi's heritage and landscape is potentially a strong constraint on a sustainable policy of development in the area. To avoid this problem, a proper approach adopted at many UNESCO sites is to develop research and studies providing guidelines for correct interventions in the fields of restoration, protection and the construction of new buildings.

At Hampi Bazaar the owners of properties such as shops, hostels, bars and houses are innovative in finding solutions for expanding their buildings. Until today no major changes could be found in the use and exploitation of the site by the local community. Apart from some uncontrolled trends in the use of traditional spaces for tourism and the illegal construction of houses (for tourist reasons), the community is still traditionally attached to the built heritage. In order to keep the authenticity of all 1,600 listed ancient structures, especially at Hampi Bazaar, the community will be encouraged to participate in a voluntary maintenance programme consisting of basic upkeep and monitoring activities. This programme is aimed at providing cultural incentives for the local community to become involved more actively in protecting sites, and at spreading the knowledge of rules and restrictions on self-restoration not only of one's own property but also of culturally relevant public edifices. The participatory approach to conservation decision making and the actual collective work for protection benefits not only the conservation status of the WHS but also the social development of the local society. Conservation activities can be used as a way to attract people from the community interested in developing technical skills.[10]

The activity of social restoration represents an effective and economic solution that is advantageous in areas with limited financial resources. The involvement of local communities in conservation processes ensures an empowerment of the cultural identity, commitment to proper protection of heritage and more individual financial investment in cultural activities. In this sense, the protection of built heritage leads also to the conservation of intangible heritage due to the long-term possibility of using edifices for cultural activities. Conservation[11] can be defined simply as any action taken to safeguard the material integrity of the cultural heritage in order to respect its significance in all aspects (cultural, artistic, historical, religious, etc.).

Public information (in the sense of publication, data management, presentation and interpretation of open and completed projects) is an important method for promoting the cultural heritage and, as a consequence, the understanding of the need for its conservation and protection. Also, the presentation of conservation projects can raise interest and encourage public participation in conservation issues. Community involvement in conservation reflects all aspects of the LCH concept such as empowerment, ownership, awareness-raising of the community's heritage, participatory development and communication between partners.

Community involvement in site interpretation

Site interpretation[12] in touristic terms is defined as the explanation of natural, cultural or historic values attached to places. It enables visitors to gain insights and understanding about the reasons for conservation and ongoing protection of heritage. It plays a critical role in achieving sustainability that leads to social benefits by empowering the participants through the development of skills in preparing interpretation projects and in communicating with visitors. Natural heritage interpretation leads to environmental benefits due to the increasing appreciation of natural and cultural landscapes and to the broader understanding of environmental issues. Through enhancing the quality of the experience of visitors and encouraging continued visitor interest, it benefits the local economy and creates economic sustainability.

Heritage interpretation is connected not only with touristic strategies but also with important issues related to the idea of citizenship. It is a way of building bridges between different marginalized groups in a community through shared involvement in working towards a common goal. This can result in the enhancement of mutual understanding and appreciation of differences and commonalities. The involvement of locals in site interpretation also assists in improving the educational level of the community. It contributes to better understanding of local history and culture, communication skills and a sense of leadership. Local guides and interpreters become experts in their professional guild, which is appreciated by national and international tourists. A higher knowledge level about local cultural activities positively impacts the safeguarding of sites and their traditions. At Hampi this instantly influences the reconnection of tangible and intangible heritage, which is important due to the strong links between cultural and natural heritage.

Interpretation strategy

The interpretation strategy for Hampi is based on a long-term mission; its three main pillars focus on site research, presentation and benefits for the local community.

The creation of a detailed data-base including all information about the natural and cultural attractions of HWHA is necessary. It is supposed to contain explicit information about every structure or building, natural formation and agricultural use of HWHA, as well as all kinds of significant cultural peculiarities concerning religion, tradition and daily life activities in the past and present. The data-base will be formed through interpretation studies and research consisting of the creation of linkages among the documentation of past interventions, current implementations and material research into both tangible and intangible heritage. The main objectives of this project are the consolidation of previous interpretation studies and a decent organization of information about tangible and intangible heritage.

The results of this research will lead to the development of a holistic understanding of the site as a cultural landscape and help in future re-nomination processes, as well as to informed decision making in creating adequate interpretation tools without threatening the outstanding universal value of HWHA.

The implementation of the data-base is an important source for using different types of high-tech information tools at HWHA, including audio guides, cultural information panels, an interactive internet web page and radio frequency identification. Besides these modern tools, traditional interpretation material will be used more extensively, including personal guides, books, workshops, informative events and training programmes.

The development of an interpretation centre at the main entrance to Hampi is considered one of the most important investments in order to stimulate tourism participation and have a strong impact on visitors. The main objectives of the interpretation centre are to give a wider idea of the importance and significance of Hampi's cultural heritage as well as to systematically integrate information services for tourists.

To achieve a high level of local integration into interpretation jobs the strategy recommends special interpretation courses for locals. The aim of the courses is to improve the participants' specific knowledge of not only the local cultural heritage but also how to present this information to foreign tourist groups. This training will allow locals to find employment with cultural tourism operators or to run their own enterprises.[13]

Conclusion

Currently, the active usage of monuments and sights represents one of the major strengths at Hampi. Locals are physically attached to their tangible heritage, which is fundamental for the continuing practice of rituals, religious manifestations and cultural activities. The presence of the inhabitants is a long-term guarantee of the preservation of Hampi's tangible and intangible heritage. Therefore, locals who are living in heritage monuments should be encouraged to stay in these areas as this can be seen as a natural maintenance strategy. The success of the community involvement strategy will be a vital asset for the overall positive outcome of the Tourism Development Strategy for HWHA.

The involvement of Hampi's community in the tourism development process is based on three major activities: inclusion in local consultative committees, participation in conservation activities and involvement in the interpretation strategy. This affects not only the cultural part of the society but also the preservation of the natural environment, which is culturally constructed. Local communities and economic systems often hold the key to their survival or destruction (Richards 2000).

The analysis has proven how useful the concept of living cultural heritage is in order to assess tourism planning and management when confronted with local livelihoods and tangible and intangible heritage. The term LCH and its definition should be adopted at a wider international level as it represents the link between tangible and intangible cultural heritage. LCH is not to be seen as a substitute for UNESCO's Living Human Treasures concept but is meant to be an expansion of the complex field of cultural heritage in general. The LCH concept goes beyond the idea of including locals individually – it aims at involving the community as a whole. As Richards and Hall (2000: 2) put it, 'communities are increasingly being seen as providing the essential link between the local and the global'; the community becomes an important aspect in the development process. The LCH concept focuses on the active involvement of the community as well as on the theoretical link between tangible and intangible cultural heritage and can be applied to any Living World Heritage Site.

The question of ownership (of ancient buildings, temples, etc.) is an important factor giving the locals the feeling of being integrated on a long-term basis. Our analysis confirms Throsby's point that planning methods are 'unlikely to be successful if they do not engage the local population in the ownership of the project' (Throsby 2009: 2). Jamal and Getz (1995) argue that residents have been marginalized and subordinated to the edge of the tourism development process and that they should be included as major stakeholders in tourism planning and development. Furthermore, the continuing presence of locals in the centre of Hampi leads to the prevention of *museumification* and supports the authenticity of the area.

This study is meant to be the first step in wider research that tests the validity and significance of the LCH concept in assisting local and international tourism management. In a PhD project the hypotheses outlined in this case study of Hampi will be tested. The foreseen research agenda includes measuring residents' attitudes towards the LCH concept; the factor of the exploitation of cultural heritage; and the tourism impact on local traditions. The acceptance and awareness of the LCH concept have to be verified and represent one of the key outcomes of the study. Anthropological research methods such as participatory observation and qualitative interviews will be used. Close cooperation is expected with the UNESCO regional office in New Delhi, the University of Kannada at Hampi and officials from the regional government.

Notes

1 Documents and reports for this chapter were made available with due acknowledgement of the Department of Tourism of Karnataka and the Hampi World Heritage Area Management Authority.

2 The HWHAMA was constituted in 2003 under the HWHAMA Act of 2002. Its main activities are the regulation and coordination of intervention within the HWHA. The urban development authority is financed by the State Department of Kannada and Culture (Government of India) and had the primary goal of developing the *Master Plan* for HWHA.

3 Based on the *Master Plan for Hampi Local Planning Area*, the total population in 2001 (census) was around 60,000. The annual growth index of the local population is around 2.3 per cent.

4 Hampi Village is located in the core zone of HWHA.

5 The term 'intangbile or living heritage' is described in UNESCO's *Intangible Cultural Heritage Info Kit No. 1* (2009): *What Is Intangible Cultural Heritage?*, and in the information paper about the outcome of the Convention of 2003 CLT/CH/ITH/KIT002.

6 UNESCO's definition of Living Human Treasures: 'Living Human Treasures are persons who possess to a high degree the knowledge and skills required for performing or re-creating specific elements of the intangible cultural heritage.' The Living Human

Treasures programme was launched in 1994 and aims at encouraging Member States to grant official recognition to talented tradition bearers and practitioners, thus contributing to the transmission of their knowledge and skills to the younger generations.

7 At Hampi the surrounding crop lands are farmed in traditional ways.

8 Feilden's guidelines emphasize that conservation 'should be viewed as a strategic tool to preserve old values in a policy of human change . . . Any adaptive use of the buildings has to be in consonance with their original functions, as also their dignity' (Feilden 1989: 96).

9 Hampi World Heritage Area Management Authority Hospet (2007): *Master Plan – 2021 for Hampi Local Planning Area.*

10 Practical skills provide a resource for social or private enterprises to expand their businesses.

11 The ICOMOS (International Council on Monuments and Sites) *Guidelines for the Education and Training in the Conservation of Monuments, Ensembles and Sites* give a clear definition of conservation: 'The object of conservation is to prolong the life of cultural heritage and, if possible, to clarify the artistic and historical messages therein without the loss of authenticity and meaning. Conservation is a cultural, artistic, technical and craft activity based on humanistic and scientific studies and systematic research. Conservation must respect the cultural context' (ICOMOS 1993).

12 An internationally approved definition by Tilden of site interpretation is the following: 'Interpretation is an educational activity which aims to reveal meanings and relationships through the use of original objects, by firsthand experience, and by illustrative media, rather than simply to communicate factual information' (Tilden, 1977: 46).

13 These skills and competences are particularly requested by local tour operators in order to identify and organize cultural tourism products and packages.

References

Feilden, B. (1989) *Guidelines for Conservation: A Technical Manual*, New Delhi, India: Indian National Trust for Art and Culture Heritage.

Hampi World Heritage Area Management Authority Hospet (2007) *Master Plan – 2021 for Hampi Local Planning Area*, India: HWHAMA.

ICOMOS (1993) Guidelines for the Education and Training in the Conservation of Monuments, Ensembles and Sites, 10th Session, Colombo, Sri Lanka. Online, www.icomos.org/docs/guidelines_for_education.html (accessed 5 April 2010).

Jamal, T. and Getz, D. (1995) 'Collaboration Theory and Community Tourism Planning', *Annals of Tourism Research*, 22: 186–204.

Miura, K. (2005) 'Conservation of Living Heritage Site: A Contradiction in Terms? A Case Study of Angkor World Heritage Site', *Conservation and Management of Archaeological Sites*, 7: 3–18.

Moscardo, G. (1998) 'Interpretation and Sustainable Tourism: Functions, Examples and Principles', *Journal of Tourism Studies,* 9: 2–13.

Munjeri, D. (1995) 'Spirit of the People, Never of Heritage', paper presented at the World Heritage Center (WHC)/International Council of Monuments and Sites (ICOMOS) meeting, Harare.

Murphy, P.E. (1985) *Tourism: A Community Approach*, New York: Methuen.

Richards, G. (2000) 'Cultural Tourism', in J. Van Der Straaten and H. Briassoulis (eds.) *Tourism and the Environment*, Dordrecht: Kluwer.

Richards, G. and Hall, D. (eds) (2000) *Tourism and Sustainable Community Development*, London: Routledge.

Taylor, G. (1995) 'The Community Approach: Does It Really Work?', *Tourism Management*, 17: 487–489.

Thakur, N. (2005) *Integrated Management Plan for Hampi World Heritage Site Volume 6*, Department of Architectural Conservation, School of Planning and Architecture, New Delhi.

Thakur, N. and Gupta, S. (2008) 'Hampi: Statutory Spatial Planning Tools for Sustainable and Value-Based Development of World Heritage Site', *Journal of the Development and Research Organisation for Nature, Arts and Heritage*, 5 (Spring/Summer): 127–136.

Throsby, D. (2009) 'Tourism, Heritage and Cultural Sustainability: Three "Golden Rules"', in L. Fusco Girard and P. Nijkamp (eds) *Cultural Tourism and Sustainable Local Development*, London: Ashgate.

Tilden, F. (1977) *Interpreting Our Heritage* (third edn). Chapel Hill: North Carolina Press.

Trimarchi, M. (2004) 'Regulation, Integration and Sustainability in the Cultural Sector', *International Journal of Heritage Studies*, 10(5): 401–415.

UNESCO (2003) Brief Convention Information CLT/CH/ITH/KIT002. Online, http://unesdoc.UNESCO.org/images/0014/001412/141247e.pdf (accessed 5 May 2010).

UNESCO (2003) Convention for the Safeguarding of the Intangible Cultural Heritage. Online, www.UNESCO.org/culture/ich/index.php?pg=00006 (accessed 5 May 2010).

UNESCO (2003) Encouraging the Transmission of ICH: Living Human Treasures. Online, www.UNESCO.org/culture/ich/index.php?pg=00061&lg=EN (accessed 15 April 2010).

UNESCO (2009) Intangible Cultural Heritage Info Kit No. 1. Online, www.UNESCO.org/culture/ich/doc/src/01851-EN.pdf (accessed 10 April 2010).

World Tourism Organization (2004) 'Report of the Seminar on Cultural Tourism and Poverty Alleviation on 8 June 2004, Cambodia'. Online, http://pub.unwto.org/WebRoot/Store/Shops/Infoshop/Products/1391/1391-1.pdf (accessed 20 April 2010).

Zappino, V. (2005) 'Caribbean Tourism and Development: An Overview', ECDPM Discussion Paper on Caribbean Tourism and Development, Discussion Paper (65) June: 22–25. European Centre for Development Policy Management, Maastricht. Online, www.ecdpm.org/Web_ECDPM/Web/Content/Download.nsf/0/C8C6E56F7D60C182C125703C004F1873/$FILE/05-65e-VZ%20-Caribbean.pdf (accessed 15 April 2011).

10 Reconstructing biodiversity for tourism development

Ethnographic accounts from a World Heritage Site in the making

Carsten Wergin

Construire la nature et ne pas la détruire![1]

There are more than 1000 World Heritage Sites (WHS) listed by the United Nations Educational, Scientific and Cultural Organisation (UNESCO). This chapter is about a site whose developers aim for it to become one. It will present parts of the particular rules and regulations as well as discuss the discursive practices that led to the reinvention of Rodrigues Island, a dependency of the Republic of Mauritius in the southwest Indian Ocean, for sustainable development based on tourist promotion.

After a brief presentation of the island, its past, present and future as a tourist destination, I first outline the motivations of local and international actors to develop it. This is accompanied by a presentation of their personal backgrounds and socio-political affiliations. The second part of the chapter deals with the effects the development project has had on local people and culture. It takes as an example the setting up of a Marine Protected Area (MPA) in the southeast of the island's 240 km^2 lagoon. I will discuss this implementation process as a 'reinvention of biodiversity' justified by globally proclaimed insecurities in contemporary society (economic, social, political) that are to be overcome through sustainable development based on heritage preservation.

In any postcolonial society, heritage is a complex and contested topic. It involves indigenous, imperial and migrant components. Rodrigues Island is no exception. In fact, the situation is compounded by the unique characteristics of the island's natural environment, the considerable isolation from other countries and its ethnic differences from its 'mother island' Mauritius. Furthermore, Rodrigues Island is a hybrid cultural space. The island has no native population, but is based on migration, deportation, exile and slavery. Most importantly, today there is a wide range of political and economic interests at play since it received the official status of autonomy within the Republic of Mauritius in 2001.

The general argument I make is for a more 'grass-roots' oriented approach to the study of World Heritage Sites in order to understand their particular importance for the larger socio-geographical context they are in from an actor's point of view. The people I engage with in this chapter are mainly local people, who often, if not always, see heritage as a means to achieve further ends than mere nature and culture preservation under UNESCO conventions. At the same time, they might be global experts for sustainable development and heritage preservation, business owners or trainers, active on various levels and in transregional contexts. Therefore, in the following pages, I refrain from further differentiation between local and global actors, as most are both at the same time.

Today the ownership of a WHS has become an important sign of being in the world. In my analysis, I therefore aim to account for heritage not as something static, but as a profound resource for development and socio-political change. Heritage is something recurring under particular rules, regulations, demands; it is a continuous cycle, from birth through evolution to death and rebirth, of 'a past tailored to present-day purposes' (Lowenthal 1998: x). The metaphor of birth and rebirth evoked here leads to a certain form of heritage that will be one focus of this text. Biodiversity is of particular significance in a more inclusive approach to heritage. The preservation of unique natural beauty and endemic species and plants is always considered of universal value by UNESCO and as such forms the basis for many successful applications to its Heritage List.

In conjunction with tourism and development, a focus on biodiversity has two additional advantages. First, many so-called 'natural' heritage sites can be found in southern regions of the globe. Consequently, their inscription turns the UNESCO project, most of which still centres around sites in the northern hemisphere, into a more global one. Second, many people from the north like to spend their holidays in beautifully preserved landscapes with at least three of the five typical tourist 'Ss' of sea, sun, sex, sand, surgery. Since preservation is costly, this is a significant step toward the financial resources required to keep biodiversity intact. The questions I will tackle in this chapter are therefore as follows: How is the preservation of Rodriguan biodiversity done? How is it fostered, and how does it help or hinder development by local people?

Heritage preservation and tourism development on Rodrigues

Heritage has been described not as an inquiry into the past but as 'a celebration of it, not an effort to know what actually happened but a profession of faith' (Lowenthal 1998: x). Such a 'Cult of Heritage' (ibid.) turns heritage foremost into a process through which the landscape and culture of a region are evaluated. International organisations send experts to help people develop their economy, preserve their landscape and culture, and lead a more sustainable life. In this context, the UNESCO World Heritage Programme has brought forth an opportunity to promote places such as Rodrigues as tourist destinations with particular natural and cultural attractions. Since its landmark Convention Concerning the Protection of the World Cultural and Natural Heritage (UNESCO 1972) the understanding of culture as a concept and its importance for the construction of social identity has both broadened and deepened considerably. It has long been pointed out, however, that if tourism is used as the financial donor, there is a danger that the richness, complexities and contestabilities of a culture will be reduced to an oversimplified set of signs and signifiers (Cohen 1988; MacCannell 1999 [1976]; Ringer 1998; Smith 1989 [1977]).

More recently, the global tourism sector has significantly changed its attitudes to the cultural resources and communities it depends upon. Due to the increased segmentation of the market, it is far more willing to engage in the sustainable development agenda (Bruner 2005; Cole 2007; Harrison and Hitchcock 2005; Williams 2004). As I will show, such market segmentation is of particular importance for places like Rodrigues, whose infrastructure and environmental and cultural fragility can only support a very limited numbers of tourists. Its local political authority, the Rodrigues Regional Assembly (RRA), has outlined various measures to overcome these obstacles, as this quotation from a speech by its then Chief Commissioner highlights:

> Our objective is to . . . attain a figure of 100,000 tourist arrivals. My Executive Council will come forward with a master plan outlining the long term policy for sustained tourism development with emphasis on the conservation of our fragile eco-systems and the island's authenticity. We shall therefore set ourselves on attracting foreign investment in this sector, aggressive marketing to achieve optimum growth, and design the regulatory framework to further those objectives.[2]

The RRA aims at a maximum of 100,000 tourists per year. This would mean that at any given moment, not taking into account high and low tourist seasons, there would be an average of 274 tourists on the island, about 0.5 per cent of the total population of roughly 35,000 (RRA-Magazine 2010). This might seem insignificant. However, it equals the respective number of Muslim, Buddhist and Hindu inhabitants of the island. Given that Mauritius puts strong emphasis on its appeal as a multi-ethnic society and terms itself the 'Rainbow nation', this is of importance, especially since figures for how many tourists are to be welcomed on Rodrigues in the longer term have been raised recently.

When landing on Rodrigues, the first surprise is the new airport, opened in 2003. It is small but light, with a beautiful view out onto the landscape. Cows and sheep are grazing within a 200 m distance. After entering the baggage claim area,

Table 10.1 Targeted 'maximum number of tourists per day in Rodrigues in selected months'

Selected months	*2003*	*2013*	*2017*	*2020*	*2025*
December	647	957	1303	1641	2411
August	548	810	1102	1389	2040
November	498	737	1002	1262	1855
January and April	448	663	902	1136	1669
Average (at any point in time during the year)	415	614	835	1052	1546

Source: Adapted from RRA-SIDPR (2009: 294).

the feeling is not of having arrived in a tourist destination but in some secluded corner of the world. Apart from a small shelf that displays artisanal products made by CARE-Co Rodrigues, a cooperative that employs disabled people, there is no advertising. Even the duty-free shop invites you less to buy than to chat with the sales person. Those who live on Rodrigues and exit the plane immediately start conversations with the ground personnel. They all seem to know each other by their first name. If they do not, the men call each other 'cousin' without any necessary family relation.

A driver welcomes most tourists on arrival and takes them directly to their hotel, following the main road that cuts through the island from southwest to northeast. For about eight kilometres they take numerous curves and climb almost to Rodrigues's highest point, Mt. Limon, at roughly 360m. The speed limit is 50 km/h and that is already pushing your luck on the winding road. In the little villages, people on either side of the road sell fruit, vegetables or grilled chicken and sausages. Others might stop their motorcycle for a quick chat with neighbours and friends.

Local politicians are well aware of the appeal that the Rodriguan lifestyle has for tourists. In the 'master plan' mentioned in the above quotation and published in 2009 under the title Sustainable Integrated Development Plan for Rodrigues (RRA-SIDPR 2009), the RRA outlines tourism as the main device for development and the local 'lifestyle' as one of the island's principal tourist attractions:

> The most important attraction of the Rodrigues destination is the very traditional and peaceful nature of the lifestyles adopted by the local population. This Rodriguan specificity, through the people being placed at the centre of the overall promotion package, is viewed as very appealing principally to people preferring a simple, non-materialistic way of life.
>
> (RRA-SIDPR 2009: 274)

This presents a conflict that is relevant not only for Rodrigues but for all existing and future WHSs: that between preservation and development. Preservation is appealing for tourists, development is the goal of tourism. If development is done in an unsustainable manner, the attraction for tourists is put at risk. Consequently, tourism might diminish, and development with it. However, looking at the proposed measures to expand Rodrigues airport, the lifestyle valorised in the aforementioned quotation might not be as important as an increase in arrival figures. In their 2010 memorandum to the newly elected government of Mauritius, the RRA proposed to:

(i) enable the landing of larger planes;
(ii) promote regional flights in Rodrigues;
(iii) attract more tourists to Rodrigues; and
(iv) encourage many of the prospective promoters to invest in their hotel development projects.

(RRA-Memorandum 2010: 2)

Parallel to this proposal, the RRA in its SIDPR identifies 'sustainable destination management' as central (RRA-SIDPR 2009: 264). In line with the above, they continue:

> The Tourism Development Plan for Rodrigues in 2002 suggested a figure of 100,000 as the carrying capacity (to be reached by 2015) but a figure of 121,000 is also included in the forecasts (to be reached by 2020) . . . It is assumed though that beyond 120,000 visitors in any one year, pressures on the environment would be intense.
>
> (RRA-SIDPR 2009: 284;, see also Table 10.1)

On Rodrigues, the development of tourism infrastructure is further complicated by the island's aforementioned mountainous landscape. Flat areas are limited and so is the space for larger hotels. The construction of an all-inclusive resort, as on its mother island Mauritius, would significantly affect the cultural landscape of the island. Taking such particularities into account, policy and planning goals for projects such as the MPA discussed below aim toward a more proactive role of the local population in the development process.

Problems that Rodrigues has been facing for a long time are unemployment and underdevelopment. The economic resources apart from tourism are limited to fishing, agriculture and day-to-day businesses. Main incomes, apart from the social benefits from the Mauritian authorities, are generated through government jobs or those in the education sector. To change this, international interventions such as the United Nations Development Programme (UNDP) or the Decentralised Development Programme (DCP), established in a collaboration of the Mauritian authorities with the European Union (EU), have brought training and financial support for the start-up phases of diverse development projects. People from numerous entities are involved in these. They include Rodriguan politicians, teachers and entrepreneurs, but also experts sent through the aforementioned initiatives and other people normally based outside the island, notably on Mauritius.

Any such agency is not naturally given but something that these experts acquire through their interaction with others. While they re-create the culture and identity of Rodrigues, emphasis also needs to be put on the significance of the local population, which only appears as passive receiver. In fact, it has the power to give agency. As will be shown in the analysis below, its patiency, the 'significant other' of agency (Schnepel 2009), brings to the fore 'hidden relations of production' inherent to the tourism industry and sustainable development projects (Tsing 2000; with particular reference to tourism, see Wergin and Neveling 2009).

Case study: a sustainable development project and its hidden relations of production

The argument so far has shown how diversity and a traditional lifestyle become significant 'earners' in tourism development initiatives. The problem that arises is that over-sensitivity to local knowledge can block people from scientific, technological or other benefits in the name of a false morality that proclaims development through preservation (Butcher 2003). Crudely speaking, it might demand of people to present themselves as 'frozen in time' and 'unspoiled by development'.[3] In line with the above, the following case study presents material collected during a decisive phase in the implementation process of a project that aimed to reshape Rodriguan landscape, nature, culture and people.

The economy of the Republic of Mauritius, being a Small Island Developing State, is heavily dependent on its marine ecosystem and biodiversity. The National Environment Strategies (NES) and the National Environmental Action Plan (NEAP) II, published in 1999, recognise the preservation of the coastal zone and the marine biodiversity as being of the utmost importance. This is to be achieved through the establishment of Marine Protected Areas (MPAs). By 2009, Mauritius had created two MPAs, in Blue Bay and Balaclava. Both are managed through a command and control model but do not achieve the desired results. Therefore, a co-management model has been applied.

The UNDP was asked to implement a UNDP/GEF (Global Environment Facility) medium size project to assist in strengthening systemic, institutional and human resource capacity. Its name was Partnership for Marine Protected Areas in Mauritius and Rodrigues. This project obtained the support of the Government of Mauritius (GoM), the RRA and local stakeholders consulted during the project development phase. Its total value amounts to 4,293,260 US dollars.[4] The site chosen for the project lies in the southeast of Rodrigues. Its name, SEMPA (South East Marine Protected Area), had already been coined in 2005, but the project was put on hold after political power changed in the 2006 regional elections. It only became the centre of attention again in 2009, during my seven-month fieldwork phase, after the implementation of the UNDP/GEF measures.

The potential of Rodrigues for sustainable tourism development was already acknowledged in a 2001 study by the international consulting firm Deloitte & Touche. Its biodiversity – endemic birds, large caves, beaches and a 240 km^2 lagoon with tropical fishes and colourful corals – was identified as key for promoting tourism. But this diversity has suffered from cyclones, overfishing and mismanagement. An estimated 40 per cent of the lagoon's coral and associated ecosystem has been severely impacted by destructive fishing practices, siltation from upstream erosion and pollution from agricultural and sewage runoff (Schils et al. 2004). Many communities in Rodrigues are dependent on the lagoon, with some 2000 fishermen attempting to earn a living from it, 900 of whom work in the southern sector where SEMPA was initiated.

The project was set up to establish the aforementioned co-management approach, which was later to be replicated in other MPAs in Mauritius. In 2009, I participated in a two-week workshop organised by the UNDP that brought together tour operators, diving instructors, kite surfing teachers, but also fishermen and politicians. International trainers were invited to work with the local population in and around the village of Port Sud-Est. They were to collectively demarcate a zone for SEMPA of roughly 45 km^2. In addition, they were to discuss

Figure 10.1 Participants in the SEMPA workshop with the lagoon in the background, Port Sud-Est, April 2009. Photograph by the author.

what could be done to further protect biodiversity, and what should be done to help the local population find new jobs, i.e. give people a different perspective on life, based on tourist promotion and environmental protection.

As mentioned above, SEMPA was to contribute to national efforts aimed at conserving the marine biodiversity of Mauritius by developing financial mechanisms and institutional and human resource capacity for the creation of a national system of MPAs. As such, its aim went well beyond Rodriguan shores. It was to develop a general enabling policy and institutional framework to sustainably co-manage MPAs throughout the Republic of Mauritius. So the island of Rodrigues – the little 'stepchild' of Mauritius with its rebellious Catholic Creole population, so much in contrast and social conflict with the Hindu dominated Mauritian system (Wergin 2012) – was not only to become a lucrative, income generating source on the contemporary tourism market; it was also to become a role model for the mother island.

Before joining SEMPA in early 2009, one of the international experts recruited had for the previous ten years been working with MPAs in the Seychelles. UNDP chose him because of his long-standing experience in the region, but also his knowledge of Creole. Rodriguan Creole is different from that of the Seychelles. However, as I witnessed myself during workshop sessions, he was able to understand most of what was said, as well as communicate in Creole with local people.

The central problem this expert identified in a personal interview I conducted with him was that of education. Repeatedly drawing on his experience in the Seychelles, where MPAs have existed for over 35 years, he argued that Rodrigues needed environmental education in schools in order for future generations to react positively to the natural environment. Apart from this 'change of mentality', tourism also played a crucial role (Personal Interview, 10 June 2009). It would generate possibilities for SEMPA to charge fees, a source of financing future management. He again emphasised this with reference to the Seychelles, where most MPAs are self-sustaining and an ecotourism plan is in place that includes sensitisation for all visitors, 'starting at the airport' (Personal Interview, 10 June 2009).

But tourists not only come to Rodrigues to stay in a hotel. Their aim is to explore. As such, one is faced with two-way work of conservation and tourism evolving at the same. One central problem the expert identified 'is not to get greedy of having hotels that are not supposed to be built. You have to have certain limits' (Personal Interview, 10 June 2009). This links to the notion of patiency presented above. It underlines the need for a local population to take its time when making decisions both for and against apparently lucrative tourist investments that, in the long run, could endanger rather than sustain development.

An arena from which to gain the necessary know-how to take such decisions is the Indian Ocean region itself with its numerous islands and neighbouring states, such as the Seychelles, from which to draw already developed expertise. In order to do so, the expert had been travelling to various meetings over the years. When we spoke, he was about to visit La Réunion for a network meeting organised by the Indian Ocean Commission (IOC), an intergovernmental organisation formed by the Comores, Madagascar, Mauritius, La Réunion and the Seychelles. Its aim is to encourage cooperation mainly in the areas of economics and commerce, but also in the conservation of resources and ecosystems. In the expert's view, such initiatives demonstrate that the Indian Ocean countries 'all get along very well':

> But one has to be careful also. You don't want to send someone to another country who, when he comes back, is being moved to a position where he cannot apply what he has learnt. Don't choose a group of your favourite friends . . . Send people from within the MPAs and get them to do what has been planned for them.
>
> (Personal Interview, 10 June 2009)

What becomes evident is the political power-play linked to apparently beneficial projects such as SEMPA. The favouritism of giving lucrative jobs or possibilities for exclusive business trips to friends instead of the most qualified personnel is only one aspect. Another are the politicians who use projects such as SEMPA to establish themselves in front of their electorate as open-minded experts with a 'world-vision' for Rodrigues.

Two days before this interview, celebrations of World Ocean Day 2009 were held in Port Sud-Est, in front of the newly renovated SEMPA building. For the first time, I witnessed the intermingling of local politicians with the SEMPA board, a group of people that had been formed during the workshop I attended a few months before. Their main audience consisted of young children between 10 and 13 years old who had been brought from different parts of the island on buses to take part in the ceremony.

The speakers used the opportunity to tell the schoolchildren in front of different media representatives about the importance of the world's oceans, and the protection of the environment in general. SEMPA had gained a political quality or, one might say, had become an instrument usable on the political stage. A government official responsible for the marine park explained about the loss of biodiversity due to climate change and, consequently, the strong commitment of his commission to international standards. He emphasised that SEMPA would be the largest MPA in the Republic of Mauritius, with fishermen and tourist stakeholders working together on its executive board. The Chief Commissioner per Interim followed him on the podium and proclaimed that since the Rio Convention in 1992 in particular the larger countries had not done enough 'to save the planet'. And even with the Kyoto Protocol the problem would grow rapidly. 'This is why it is great that Rodrigues, even as it is a very small island, does something.'

After the speeches had ended, one member of the SEMPA board took me aside to emphasise that it still remained to be seen how, despite this momentary success, there would be enough money available to run SEMPA after UNDP/GEF had withdrawn from the project. By then, rules and regulations, for example about the licensing of the hotels, fishermen and tourist agencies through which SEMPA would be financed, were meant to be in place, cross-signed by the Mauritian and Rodriguan authorities. But, for him, the local government seemed more interested in having 'the final word' itself than in giving the newly elected SEMPA board the necessary powers.

Conclusion

The Chief Commissioner per Interim, in his speech on World Ocean Day 2009, summed up the task of SEMPA in a simple phrase: 'Construire la nature et ne pas la détruire!' The slogan went down well with his audience. There was no question that what they were doing at SEMPA was not preserving but constructing nature. From an analytical point of view, the missing link in the nature preservation/ tourism development conflict becomes construction. The question to be answered then is less about who constructs than who has the power to decide what is constructed.

What the above has shown is how politicians, entrepreneurs and others involved in tourist promotion draw attention to landscape and culture by employing scientifically imbued arguments with which they present a particular natural or cultural setting. Slogans such as 'SEMPA', a play on the colloquial French word for 'likeable', are used to provide a positive image of an open-minded and welcoming population that may override a history of Creole people whose ancestors, at one point in time, were forcibly relocated to a place such as Rodrigues.

SEMPA demonstrates the entanglement of World Heritage and tourism with the power geometries of everyday life. Those are as much economic development tools as they are scientific discourses about the professed value of particular material and symbolic capital (Ateljevic et al. 2007; Poria and Ashworth 2009). SEMPA is to give Rodrigues a place in the international tourism arena. For this, the local is induced with new ideas, with ideologies of sustainability and nature conservation that transform people, place, history and the tradition of the island. What becomes evident is that the place for the 'acting out' of heritage and cultural difference lies neither solely on a global nor a local level, but somewhere in between: in the interdependencies between the local and the global, in manifestations and alterations of globalist actions and discourses through the impact of patient local actors.

In more recent anthropological engagement, criticism has been raised against UNESCO and its notion of 'intangible heritage' (2003) to the point where it might create legally enforceable entitlements on the part of indigenous peoples. However, rather than such a risk of 'natural copyright' (Comaroff and Comaroff 2009: 37), the more pressing issue seems to lie in finding suitable measures of what qualifies as intangible culture and heritage in the first place (Brown 2005: 51). Analysed through its hidden relations of production, SEMPA is as much a sustainable development project as it is about controlling people and their needs, space, biodiversity, money, politics, social relations, culture. Sustainability as ideology proclaims places unspoiled by development and creates particular forms of 'zoning'. Researchers need to take such re-organisation of space into consideration. The lagoon of Rodrigues was first described by fishermen with particular names that reminded them of their long journeys outside the lagoon, as sailors on large fishing boats in the Indian Ocean. Such 'traditional zoning' of fishermen is today replaced by the SEMPA zoning, in favour of preserving the fish.

While a sustainable development project appears to bring conservationists and economists together in a united goal for 'constructive' preservation, it can also end in false lobbying and labelling that lead to a reassertion of powerful interest groups. On a local level, the divide between rich and poor might even increase, as becomes visible in the political context of small islands like Rodrigues, where a drive for sustainable development is arguably used by some to enhance their individual power. While this is not to question the measures of the UNESCO committee already in place, one has to acknowledge that a more integral approach to the evaluation of a site is needed. 'Development' needs to be understood as a multi-sited intervention.

The above has shown how local people often find themselves in a contradictory setting between development and preservation. While actors expect them to form a homogeneous group, they remain diverse in their social backgrounds, interests and goals. Researchers need to take a closer look at how the lobbying is done, and how different actors play out their interests and use the advocacy of sustainable development for their own purposes. Rodriguan lifestyle and biodiversity are turned into heritage in the very sense that Lowenthal gives to this term: 'a past tailored to present-day purposes' (as discussed above). Such heritage, not only on Rodrigues, has taken on a more instrumental and pragmatic role in social life. An analysis of the linkages between heritage preservation and sustainable development allows us to look at the tourism industry as a category that holds further possibilities within it, no matter how critical one might be of it.

Notes

1 'Construct the future, not destroy it!' Gaëtan Jhabeemissur, at the time Rodrigues Chief Commissioner per Interim, in his speech on World Ocean Day 2009 in Port Sud-Est, Rodrigues.
2 Excerpt from the address to the RRA by Chief Commissioner Johnson Roussety on 22 December 2006, outlining the Regional Government Programme for the period 2006 to 2011 (source: RRA archives).
3 In an online search for the word grouping 'unspoiled by development', conducted with the search engine Google on 14 April 2011, the first ten results out of an estimated 28,300 all used this phrase to advertise tourist destinations, from Canada to New Zealand, Dubrovnik via boat, to scuba diving in the Xcalak Reef National Marine Park on the Mexico–Belize border.
4 According to the SEMPA homepage, the sum is shared by GEF (978,000 US dollars), UNDP (28,320 US dollars), GoM/RRA co-sharing (1,302,940 US dollars), GoM in kind (208,000 US dollars) and EU parallel funding (1,776,000 US dollars).

References

Ateljevic, I., Pritchard, A. and Morgan, N. (eds.) (2007) *The Critical Turn in Tourism Studies: Innovative Research Methods*, Oxford: Elsevier.

Brown, M.F. (2005) 'Heritage Trouble: Recent Work on the Protection of Intangible Cultural Property', *International Journal of Cultural Property*, 12: 40–61.

Bruner, E.M. (2005) *Culture on Tour: Ethnographies of Travel*, Chicago, IL: Chicago University Press.

Butcher, J. (2003) *The Moralisation of Tourism: Sun, Sand . . . and Saving the World*, London: Routledge.

Cohen, E. (1988) 'Authenticity and Commoditization in Tourism', *Annals of Tourism Research*, 15: 371–386.

Cole, St. (2007) 'Beyond Authenticity and Commodification', *Annals of Tourism Research*, 34: 943–960.

Comaroff, J. and Comaroff, J. (2009) *Ethnicity Inc.*, Chicago, IL: Chicago University Press.

Harrison, D. and Hitchcock, M. (eds.) (2005) *The Politics of World Heritage: Negotiating Tourism and Conservation*, Clevedon, UK: Channel View Publications.

Lowenthal, D. (1998) *The Heritage Crusade and the Spoils of History*, Cambridge: Cambridge University Press.

MacCannell, D. (1999 [1976]) *The Tourist: A New Theory of the Leisure Class*, Berkeley: University of California Press.

Poria, Y. and Ashworth, G. (2009) 'Heritage Tourism: Current Resource for Conflict', *Annals of Tourism Research*, 36: 522–525.

Ringer, G. (ed.) (1998) *Destinations: Cultural Landscapes of Tourism*, London: Routledge.

RRA-Magazine (April 2010) Magazine of the Rodrigues Regional Assembly. Online, www.gov.mu/portal/sites/rra_portal/chiefcomm/rramagazine.htm (accessed 15 September 2010).

RRA-Memorandum (May 2010) Memorandum to the Newly Elected Government of Mauritius. Online, www.gov.mu/portal/sites/rra_portal/download/memo10.pdf (accessed 15 September 2010).

RRA-SIDPR (2009) Sustainable Integrated Development Plan for Rodrigues. Online, www.gov.mu/portal/sites/rra_portal/chiefcomm/sidpr.htm (accessed 15 September 2010).

Schils, T. et al. (2004) 'The Marine Flora of Rodrigues (Republic of Mauritius, Indian Ocean): An Island with Low Habitat Diversity or One in the Process of Colonization?', *Journal of Natural History*, 38: 3059–3076.

Schnepel, B. (2009) 'Zur Dialektik von Agency und Patiency', *Paragrana*, 18 (2): 15–22.

SEMPA Homepage. Online, www.sempa-rodrigues.com/index.php?id=2 (accessed 15 September 2010).

Smith, V. (ed.) (1989 [1977]) *Hosts and Guests: The Anthropology of Tourism*, Philadelphia: University of Pennsylvania Press.

Tsing, A. (2000) 'The Global Situation', *Cultural Anthropology*, 15: 327–60.

UNESCO Criteria. Online, http://whc.UNESCO.org/en/criteria (accessed 23 September 2010).

UNESCO (1972) Convention Concerning the Protection of the World Cultural and Natural Heritage. Online, http://whc.UNESCO.org/archive/convention-en.pdf (accessed 22 September 2010).

UNESCO (2003) Convention for the Safeguarding of the Intangible Cultural Heritage. Online, http://unesdoc.UNESCO.org/images/0013/001325/132540e.pdf (accessed 23 September 2010).

Wergin, C. (2012) 'Trumping the Ethnic Card: How Tourism Entrepreneurs on Rodrigues Tackled the 2008 Financial Crisis', *Island Studies Journal*, 7 (1): 119–134.

Wergin, C. and Neveling, P. (2009) 'Projects of Scale-Making: New Perspectives for the Anthropology of Tourism', *Etnográfica*, 13 (2): 315–342.

Williams, St. (ed.) (2004) *Tourism: Critical Concepts in the Social Sciences*, London: Routledge.

11 Post-inscription challenges

Renegotiating World Heritage management in the Laponia Area in Sweden

Patrick Brouder

World Heritage and tourism

Of the more than 1,000 United Nations Educational, Scientific and Cultural Organization (UNESCO) World Heritage Sites in the world in 2016, only 32 are classified as mixed sites, being protected for both their natural and their cultural heritage value. In the Nordic countries there are several dozen World Heritage Sites and among these the Laponia World Heritage Area stands alone as the only mixed site in the Nordic realm. The World Heritage Committee considers all sites to have 'outstanding universal value' (World Heritage Manuals 2002: 11) and the less than 4 percent of all sites which are valued for their natural and cultural value are especially worthy of study.

'World Heritage listing is clearly a significant factor for tourism management' (Hall and Piggin 2002: 103) and even though a direct relationship between listing and a tourism increase is not guaranteed (Hall and Piggin 2001), the sensitive nature of World Heritage Sites warrants careful appraisal of potential tourism impacts. In the case of mixed sites, particularly those with a 'living culture', the sensitivity to tourism impacts may be somewhat piqued. In northern Sweden the living culture of the Sami reindeer herders conjures images which are very appealing to international tourism (Pettersson 2003). Also, since the remit of the Laponia World Heritage Area is, in part, to preserve Sami reindeer husbandry, the question may be asked as to what exactly is being preserved and to what end. Is Sami reindeer herding being preserved in order to protect the heritage or to maintain and promote an iconic attraction of northern Sweden to tourists?

Certainly, the World Heritage Committee is concerned with 'assuring the conservation and protection of the World's heritage' (UNESCO 1972: 1) and not directly with tourism issues. However, a consequence of listing may be an enhanced tourism profile and so, regardless of intentions, tourism becomes a factor in 'conservation and protection', a factor which may be positive or negative but which must be managed. In the Laponia World Heritage Area tourism certainly exists and tourism to northern Sweden is growing (SCB 2010: 7). It presents opportunities for entrepreneurs in the region, while also threatening to create conflict between the local stakeholders. Whether the tourists come for nature or culture is, perhaps, less relevant than how they interact with nature and culture within the Laponia World Heritage Area. Moreover, from a supply-side perspective one may ask: who should be allowed to use the area for commercial purposes and, following this, how should commercial use be managed? Of course, tourism is not the only factor to be considered; it is, rather, one of myriad challenges facing the region as a whole and the World Heritage Area in particular. This chapter, however, will only consider the changing tourism milieu of the Laponia Area.

Aim and research questions

There is a new management structure for Laponia, which has emerged as a result of the Laponia process, a process instigated to identify the key management issues for the area, with the negotiations being completed in 2009. The new management structure involves Sami representatives, local municipalities, the county administration and the National Environmental Protection Agency. The aim of this study is to investigate how Norrbotten county's tourism businesses value the Laponia Area and to highlight for the new management the fact that the geographical reach of Laponia needs to be considered in planning.

The following research questions are addressed:

- What do Norrbotten county's tourism businesses think of the Laponia Area in particular and protected nature in general?
- What is the value of the Laponia Area's attributes (Sami culture and nature) for tourism businesses in Norrbotten county?
- What is the market reach of the Laponia Area in Norrbotten county and how may this affect management planning?

The Laponia process: a new beginning

The Laponia World Heritage Area in northern Sweden is one of only 32 places on the World Heritage list which is protected for both its natural and cultural value. The natural landscape owes its value to the receding glaciers of the last great ice age, which have left behind large marshlands, deltas and pristine forests as well as unique vegetation. Moreover, the cultural value of the region is second to none, with the Sami people being the only indigenous people in Europe whose members still practise, at least to some extent, expansive animal herding. The Sami are reindeer herders and the Laponia World Heritage Area, covering 9,400 km^2 (see map of Laponia in Figure 11.1), was created to assist the local Sami in preserving their traditions.[1]

Some of the land now making up the Laponia Area was part of the establishment of Europe's first national parks in 1909 (Mose 2007). The Laponia World Heritage Area was listed in 1996 and for the first decade or so was managed simply as a large park. However, Sami concerns over the management of the area brought to the fore the concept of 'mankind as part of nature' (Mose 2007: 23), implying that traditional park management no longer sufficed. In 2007 the Laponia delegation was charged with identifying the key management issues for the area, with the negotiations being completed in December 2009. A final report was submitted to

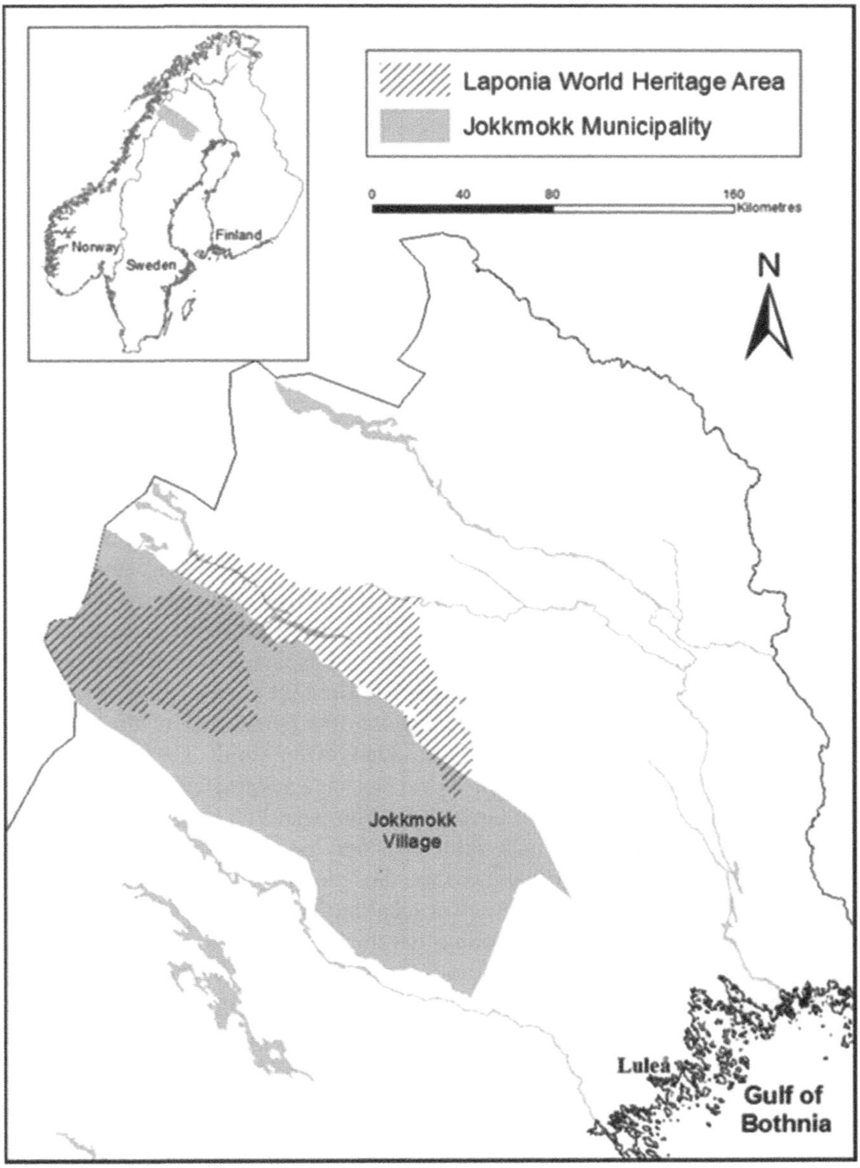

Figure 11.1 Laponia Area and Jokkmokk Municipality in Arctic Sweden.

Source: Brouder (2012).

the Swedish government in June 2010. Now the Laponia World Heritage Area has an agreed management structure involving a majority of Sami representatives, as well as the local municipalities of Jokkmokk and Gällivare, the county of

Norrbotten and the Swedish Environmental Protection Agency. The overall goal is to lead to coordinated development of the Laponia World Heritage Area. The Laponia process should not be seen as a terminal point but rather as a new beginning for the Laponia Area.

Managing World Heritage tourism in a peripheral milieu

Being a World Heritage Area

As stated above, the Laponia World Heritage Area is a mixed site being protected for both its natural and its cultural value. However, 'there would never have been a Laponian World Heritage area as we know it today, if the Saami Parliament had not objected to the Laponian area being labeled "a wilderness"' (Nilsson Dahlström 2003: 241). The proposed label of a wilderness area was contested by the Sami since: 'Archaeological remains and traces of Saami activities make the area primarily a cultural landscape and not a wilderness area (the Saami Parliament March 1995)' (ibid.: 253). The concept of 'wilderness' as a contested term has been discussed elsewhere (e.g. Saarinen 2005). For indigenous peoples, such as the Sami reindeer herders, whose way of living has evolved over centuries, 'the distinction between "nature" and "culture" is in many ways an irrelevant distinction to make' (Nilsson Dahlström 2003: 169). This distinction, however, is made and, somewhat surprisingly, the cultural aspect comes off second best: 'the fact that a large part of the Laponian World Heritage Area was and is a grazing landscape is conveniently ignored in favour of a perception of the area as a wilderness area, with a little bit of culture added on top' (ibid.: 239). However, the listing of Laponia as a mixed site of World Heritage acknowledges that the Sami culture is of special value, even beyond the European Polar region, and that the area until recently perceived as a wilderness may begin to be seen as a cultural haven. In fact Green (2009: 19) argues that 'the "global eye" of the World Heritage Organization also can work to empower local/indigenous peoples in their ethno-political objectives'. It is worth noting that the area now known as the Laponia Area was already protected land under Swedish law before 1996 and so listing is 'not something which takes place instead of the national protection, but rather something that underlies and strengthens this protection' (Nilsson Dahlström, 2003: 209). Arguably, the change brought by listing is positive since Sami culture is elevated in status. Tourism impacts, however, could be positive or negative.

Tourism and the Laponia World Heritage Area

Tourism in the Laponia World Heritage Area includes nature-based tourism and cultural tourism. Nature-based tourism is a significant segment in the north of Sweden, where a high percentage of the land is protected (24 percent of all Swedish protected land is in Norrbotten county; Lundmark and Sjernström 2009), and nature-based tourism accounts for approximately 8 percent of all tourism revenue in Sweden (Fredman et al. 2008). Cultural tourism is present within the study area and 'languages, traditions, ethnic symbols and the aboriginality status have for a long time contributed to the touristic attractiveness of the Sami core area' (Viken and Müller 2006: 1). Moreover, two major Sami attractions in the region bring in over 40,000 visitors each per annum: the Ajtte Museum in Jokkmokk (45,000 visitors) and the Jokkmokk winter festival (40,000 annually over three days), suggesting there is a palpable attraction to the cultural heritage of the local Sami (ibid.). However, despite these marquee attractions, Sami culture is often just an 'exotic backdrop for a . . . mountain vacation in the area' (ibid.: 2) rather than being a significant 'pull' factor itself for tourism in the region. The question is: can (or should) the newly established Laponia management, which has a Sami majority, elevate Sami tourism and market it more vigorously or attempt to limit tourism to the area?

Previous studies have discussed what the added value, if any, is of the World Heritage label. A study of New Zealand WHAs found that the 'national parks were attractions enough in themselves and did not require World Heritage status to make them attractive to tourists' (Hall and Piggin 2002: 406). The New Zealand entrepreneurs claimed that tourists 'come because of specific attractions – glaciers, mountains' (ibid.: 408) and were not particularly interested in the World Heritage listing. In fact, there may be a '"layer of attractions", with the World Heritage listed park constituting an attraction in its own right in terms of its capacity to draw international and domestic visitors with individual sightseeing attractions existing within the park' (ibid.: 402). Therefore it may be that a given World Heritage Area's ability to attract tourists is a function of the quality and number of the attractions to be found within its limits and the prestige attributed to the title of World Heritage. Similarly, Lundmark and Sjernström (2009: 397) claim that a 'national park in general does not itself constitute the main reason for visiting . . . although it could be a contributing factor'. On the other hand, 'Heightened awareness . . . leads to greater consideration and a general rise in the level of protection and conservation afforded to heritage' (Fyall et al. 2003: 208) and it is worth considering this symbolic 'layer of protection' which comes with listing.

Tourism development and changing management

When it comes to evaluating the potential of World Heritage listing for regional development, researchers have faced the classic preservation versus development conundrum. In Sweden, this has traditionally meant preservation, possibly at the expense of tourism development. However, the 'Swedish Environmental Protection Agency is now considering reducing the obstacles affecting commercial activities in the national parks' (Lundmark and Sjernström 2009: 402). In the case of the Laponia Area, 'it is claimed by the Sami that the Sami themselves should control tourism development to guarantee an authentic tourist experience and a survival of Sami culture' (Müller and Pettersson 2006: 55). The Sami should have control of tourism in the Laponia Area since it is their reindeer herders who form the 'exotic backdrop' and since it is their culture being put on display. Indeed, there needs to be a negotiation between the Sami (what they are willing to offer) and the tourists (what they wish to experience) (ibid.). Moreover, Pettersson (2003) examined the power relations between Sami and non-Sami stakeholders at the winter festival. The winter festival, organized by Jokkmokk Municipality, gives the balance of power to the Sami interests, but what of the balance of power in year-round tourism to the Laponia Area which is spread out across a vast territory?

In any case, tourism development is occurring and will continue to result in change for the Sami people within the region since 'even a Sami tourism development would lead to a situation where a lot of Sami are exposed to tourists whether they like it or not' (Müller and Pettersson 2006: 60). If tourism development is inevitable, then it may be preferable for the Laponia management to take a more proactive approach in order to keep the balance of power in their favour. The current management situation inevitably leads to consideration of how much development is desired or even possible (Brouder 2012). Hammer notes that to move from a lower level boost to a situation of significant development requires 'an appropriate strategic course of action' (Hammer 2007: 28). The challenge of operationalizing a strategy for the Laponia Area is discussed below.

Moving from Laponia designation to Laponia destination management

Amin (1999) identified a key endogenous growth strategy of building clusters and local economies of association, yet in the case of Sami tourism this has been 'constrained by the small scale of Sami tourism in the area and the difficulties in allying with non-indigenous stakeholders' (Müller and Kuoljok Huuva 2009: 125). From a destination development perspective, there may be a significant revenue loss if, as Müller and Kuoljok Huuva state, 'it seemed that an existing demand sometimes had difficulties in actually finding the Sami' (ibid.: 124). There are a number of contextual challenges to galvanizing growth in areas such as Laponia: lack of social agreement; increasing calls for democratization of governance; declining capacity of institutions; and diversifying populations in or around the region (Bushell and Eagles 2007). While the first three points on this list have, arguably, been addressed by the Laponia process, the challenge for the new management may well be the actions of the population surrounding the area, i.e. can a new network or local economy of association position itself to capitalize on the possible latent demand for Sami cultural tourism? This is by no means an easy task since cooperation 'required time and regular contact, which it was sometimes difficult to maintain' (Müller and Kuoljok Huuva, 2009: 124). The new management has a fresh chance to move the process beyond designation negotiation to destination management.

Being a peripheral destination

McKercher (1998: 47) has written that 'areas with poorer market access rely less on primary built attractions, trading more on the intrinsic appeal of the area'.

Elsewhere, Boyd (2002) has shown that over one third of trips at a distance of over 800 km have culture as a primary motivation. So it may be the case that cultural tourists are willing to travel further, but there must be a tangible reason for them to travel. Thus the potential exists to promote the Laponia Area utilizing both the attraction of Sami culture and the attraction of the natural environment. Of course, negotiating the cross-cultural relationships raises the question of tourist expectations vis-à-vis local offers (Hinch and Delamere 1993) but the peripherality of the Laponia Area means that while it is sheltered from mass tourism, it is also required to make a concerted, unified, community effort in order to develop tourism.

Method: surveying the periphery

A questionnaire was sent in December 2009 to a sample of 140 tourism businesses operating in Norrbotten county using an online survey resource (surveymonkey.com), which resulted in 77 usable responses (55 percent), with 69 respondents (49 percent) completing the survey. The breakdown of tourism businesses by geographical location (contiguous with Laponia Area or non-contiguous with Laponia Area) was approximately a 10/90 split. From the reporting municipalities, 7 cases were contiguous with the Laponia Area and 62 cases were non-contiguous, with continguity here defined as businesses operating from Jokkmokk or Gällivare Municipality.

Fisher's Exact Test was used in the analysis, with the test scores presented, where appropriate, in the results section. 'Fisher's test is used to determine if two groups differ in the proportion with which they fall into two categories. It is a particularly useful nonparametric technique for examining discrete ordinal or nominal data when two independent samples are small in size' (Coshall 1989: 38). The chosen level of significance was 0.05 for all non-parametric testing. Likert-scale results are reported in the text for descriptive purposes and the non-parametric tests appear in Tables 11.1, 11.2 and 11.3.

A failure analysis was performed to consider the non-respondents: 30 replied to an email and stated either they did not have time to answer the online survey (50 percent) or they felt they were too far away from the Laponia Area for it to matter to them (17 percent), with the remainder (33 percent) stating that their internet connection was too slow for the online survey.

Table 11.1 Value of Laponia Area values for businesses

	Sami valued	*Nature valued*
Cultural : Other	**89% : 46%** *(n = 9 : 68)* *Fisher's Exact Test = 0.016*	**100% : 97%** *(n = 9 : 68)* *Fisher's Exact Test = 0.198*
Contiguous : Non-contiguous	**75% : 48%** *(n = 8 : 69)* *Fisher's Exact Test = 0.14*	**88% : 99%** *(n = 8 : 69)* *Fisher's Exact Test = 0.779*

Note: n = 77; survey asked how Sami culture and nature are valued only from a business perspective.

Table 11.2 Laponia Area has value for respondent's business (location)

Location	*Yes*	*No / Don't Know*	*Total*
Contiguous	5	2	7
Non-contiguous	19	43	62
Total	**24**	**45**	**69**

Notes: Fisher's Exact Test = 0.045; also, Laponia Area is valued by customers: Fisher's Exact Test = 0.019.

Table 11.3 Laponia Area has value for respondent's business (main product)

Main Product	*Yes*	*No / Don't Know*	*Total*
Cultural tourism	5	2	7
Other tourism	19	43	62
Total	**24**	**45**	**69**

Notes: Fisher's Exact Test = 0.045; also, Laponia Area is valued by customers: Fisher's Exact Test = 0.019.

Result: nearness matters but attractiveness overcomes distance

Descriptive results show the majority of businesses agree that the Laponia Area is good for the county of Norrbotten and feel also that it is good for the interior region of Lapland. This is not a surprising result but demonstrates that tourism businesses are positively disposed to the World Heritage listing. When asked about national parks and tourism development, the majority stated that national parks do not hinder development but there was quite a large deviation in the sample. Furthermore, when asked if protected nature should be open for tourism, the majority answered positively. Taken together these results may exhibit a reluctance to commit to the idea of national parks unless access for tourism businesses is allowed, as is the case in the Laponia Area, and then tourism businesses can accept that the protected area is positive for the region. Finally, the majority of businesses would also like to see the development of a Lapland tourism network in order to aid the growth of tourism in the county. There is no existing network of all Lapland tourism businesses, and an opportunity exists for the new Laponia management to meet this latent demand. A certain caveat here is that even the strongest results are not absolutely strong but rather relatively strong. This can be explained somewhat by the geography of the area. It is expansive and it may well be that the average tourism entrepreneur is not very excited about, or interested in, the Laponia Area. However, a closer look at the differences within groups offers a more detailed picture of the value of the Laponia Area.

Table 11.1 displays the views of businesses with regard to the two core values of the Laponia Area (Sami culture and nature). The results are divided into two separate categories for comparison – cultural tourism businesses vis-à-vis other tourism businesses and contiguous vis-à-vis non-contiguous municipalities, the idea being that cultural tourism businesses may value Sami culture more, regardless of location, and that, possibly, businesses in contiguous municipalities may have greater emphasis on nature since the Laponia Area is protected for its particularly unique natural assets. The results, however, show that the only statistically significant outcome is that cultural tourism businesses value Sami culture more than other businesses. Within contiguous municipalities, businesses place a stronger value on Sami culture than they do in non-contiguous municipalities, although this result is not statistically significant (75% : 48%). Examining the 'Nature valued' column, the value of nature to tourism businesses is near total. This is true regardless of the type of business or location relative to protected nature and suggests that the defining feature of tourism in the north of Sweden is contact with nature, i.e. all tourism is nature-based tourism.

Table 11.2 shows one truism of market reach – proximity to the attraction matters. Here, nearness is defined as those businesses operating in two municipalities contiguous with the Laponia Area (Jokkmokk and Gällivare), referred to here as the contiguous businesses, and businesses in all other municipalities are considered non-contiguous (average drive time from the non-contiguous municipal centres is 2.5 hours). The results in Table 11.2 show that, as expected, there is a significantly greater proportion of contiguous businesses that value the Laponia Area (Fisher's Exact Test = 0.045) and this result is supported by another demonstrating that businesses believe their customers value the Laponia Area (Fisher's Exact Test = 0.019). As expected, the market reach of the Laponia World Heritage Area for a general sample of tourism businesses in Norrbotten county is limited to those in the contiguous municipalities. This, however, is not the complete picture.

Table 11.3 shows an interesting result. Cultural tourism businesses, regardless of their location, tend to value the Laponia Area more than other tourism businesses (Fisher's Exact Test = 0.045) and their customers also value the Laponia Area more (Fisher's Exact Test = 0.019). This is especially interesting considering the small sample of cultural tourism businesses is spread out across the county (one contiguous; six non-contiguous) and suggests that the cultural value of the Laponia Area allows its market reach to increase for the businesses involved in cultural tourism, with the cultural tourism businesses included here mostly being tour guides based on the Bothnian coast.

Discussion: how 'local' is local?

If the Laponia values are examined (Sami culture and nature) across the sample, it is seen that nature is valued by almost all, whereas Sami culture is very much place or sector dependent. This is not surprising from a business perspective since nature is a key factor in the county's attractiveness, whereas Sami tourism is somewhat of a niche segment in terms of direct business. What is not measured here, however, is the value of Sami culture as the 'exotic backdrop' discussed by Viken and Müller (2006). If it is true that Sami culture is valuable for promotion, then it may be posited that, regardless of how many businesses in the region value Sami culture, all may benefit from the positive image of Sami culture.

Müller and Kuoljok Huuva (2009) have shown that Sami tourism entrepreneurs had some difficulties connecting with enough of the other stakeholders often enough to make it worthwhile, yet the respondents across the county also strongly maintain that a Lapland network would be good for development. It remains to be seen how the new management structure will incorporate other business interests from, for example, Luleå (the county capital city on the Bothnian coast) in the tourism development of the Laponia Area.

Tables 11.2 and 11.3 show some measure of the supply-side market reach of the Laponia World Heritage Area. As expected, contiguous businesses derive more value from the Laponia Area. This is, of course, intuitive but when the focus of analysis is shifted to product offer, it is shown that cultural businesses, regardless of location, derive value from the Laponia Area. In short, nearness matters but sensitivity to distance is eased by a product match. When one considers that there are cultural businesses coming from outside the Laponia Area, it is obvious that managing the development of tourism in the Laponia Area requires not just a local focus but also a regional one, although the balance of power should remain with the local Sami and those businesses within the Laponia World Heritage Area, as agreed during the Laponia process.

Conclusion: not a terminal point but a new beginning

Tourism planners may be forgiven for examining the general sample of businesses and concluding that the Laponia Area only matters to local businesses. What this chapter demonstrates is that there is a small but significant group of businesses geographically outside the Laponia Area but, nonetheless, falling within the sphere of influence of the new Laponia management. The onus is on the Laponia management to decide how to deal with this group. One suggestion emerging from the Laponia delegation's report to the Swedish government is a new permit system for business users in the Laponia Area. This would allow better knowledge of who exactly is using the Laponia Area as well as provide an opportunity to maintain direct contact in a formal network and establish certain professional guidelines for commercial use. This case also carries a message for World Heritage Area researchers and managers alike to reappraise the often parochial views on which businesses value and utilize WHAs – changing the focus from a local to a regional level at least. Finally, in the case of Laponia, an optimal level of commercialization may be achieved through a negotiated compromise on what the local Sami want to offer and what the contiguous and non-contiguous tourist enterprises' customers want to experience. A Laponia network is an appropriate forum for learning about each other's positions and moving forward together. Therefore, the Laponia process is not a terminal point but a new beginning.

Note

1 www.laponia.nu.

References

Amin, A. (1999) 'An Institutional Perspective on Regional Development', *International Journal of Urban and Regional Development*, 23: 365–378.

Boyd, S. (2002) 'Cultural and Heritage Tourism in Canada: Opportunities, Principles and Challenges', *Tourism and Hospitality Research*, 3 (3): 211–233.

Brouder, P. (2012) 'Creative Outposts: Tourism's Place in Rural Innovation', *Tourism Planning and Development*, 9 (4): 383–396.

Bushell, R. and Eagles, P.F.J. (2007) *Tourism and Protected Areas: Benefits beyond Boundaries*, Wallingford: CABI.

Coshall, J. (1989) 'The Application of Nonparametric Statistical Tests in Geography Quantitative Methods Research Group (UK)', CATMOG 50, Norwich: University of East Anglia.

Fredman, P., Boman, M., Lundmark, L. and Mattsson, L. (2008) 'Friluftslivets ekonomiska värden – en översikt', *Rapport Friluftsliv*. Online, http://svensktfriluftsliv.se/wp-content/uploads/2012/12/Friluftslivets-ekonomiska-v%C3%A4rden-Rapport-2008.pdf (accessed 22 July 2016).

Fyall, A., Garrod, B. and Leask, A. (2003) *Managing Visitor Attractions: New Directions*, Oxford: Butterworth-Heinemann.

Green, C. (2009) 'Managing Laponia: A World Heritage Site as Arena for Sami Ethno-politics in Sweden', Uppsala Studies in Cultural Anthropologies 47. Uppsala: Acta Universitatis Uppsaliensis. Online, https://uu.diva-portal.org/smash/get/diva2:275592/FULLTEXT01.pdf.

Hall, C.M. and Piggin, R. (2001) 'Tourism and World Heritage in OECD Countries', *Tourism Recreation Research*, 26 (1): 103–105.

Hall, C.M. and Piggin, R. (2002) 'Tourism Business Knowledge of World Heritage Sites: A New Zealand Case Study', *International Journal of Tourism Research*, 4: 401–411.

Hammer, T. (2007) 'Protected Areas and Regional Development: Conflicts and Opportunities', in I. Mose (ed.) *Protected Areas and Regional Development in Europe: Towards a New Model for the 21st Century*, London: Ashgate.

Hinch, T. and Delamere, T. (1993) 'Native Festivals as Tourism Attractions: A Community Challenge', *Journal of Applied Recreation Research*, Ontario Research Council on Leisure, 18 (2): 131–142.

Laponia World Heritage (2010) Online, www.laponia.nu (accessed 1 June 2010).

Lundmark, L. and Sjernström, O. (2009) 'Environmental Protection: An Instrument for Regional Development? National Ambitions versus Local Realities in the Case of Tourism', *Scandinavian Journal of Hospitality and Tourism*, 9 (4): 387–405.

McKercher, B. (1998) 'The Effect of Market Access on Destination Choice', *Journal of Travel Research*, 37: 39–47.

Mose, I. (ed.) (2007) *Protected Areas and Regional Development in Europe: Towards a New Model for the 21st Century*, London: Ashgate.

Müller, D.K. and Kuoljok Huuva, S. (2009) 'Limits to Sami Tourism Development: The Case of Jokkmokk, Sweden', *Journal of Ecotourism*, 8 (2): 115–127.

Müller, D.K. and Pettersson, R. (2006) 'Sami Heritage at the Winter Festival in Jokkmokk, Sweden', *Scandinavian Journal of Hospitality and Tourism*, 6 (1): 54–69.

Nilsson Dahlström, Å. (2003) *Negotiating Wilderness in a Cultural Landscape*, Uppsala: University Library.

Pettersson, R. (2003) '*Indigenous Cultural Events*: The Development of a Sami Winter Festival in Northern Sweden', *Tourism*, 51 (3): 319–332.

Saarinen, J. (2005) 'Tourism in the Northern Wilderness', in C.M. Hall and S. Boyd (eds.) *Nature-Based Tourism in Peripheral Areas*, Clevedon, UK: Channel View Publications.

SCB (Structural Business Statistics) (2010) *Serie NV – Näringsverksamhet*, NV 41 SM1006, 5 May 2010, p.15.

UNESCO (1972) Convention Concerning the Protection of theWorld Cultural and Natural Heritage. Online, http://whc.UNESCO.org/en/conventiontext (accessed 21 December 2012).

Viken, A. and Müller, D.K. (2006) 'Introduction: Tourism and the Sami', *Scandinavian Journal of Hospitality and Tourism*, 6 (1): 1–6.

World Heritage Manuals (2002) '*Managing Tourism at World Heritage Sites*, No. 1, Paris: UNESCO World Heritage Centre.

12 The level of societal reproduction as a predictor of visitation

Lessons from World Heritage Sites in the United States

Linda Joyce Forristal

Since the United Nations Educational, Scientific and Cultural Organization (UNESCO) adopted the Convention Concerning the Protection of World Natural and Cultural Heritage, commonly referred to as the World Heritage Convention (UNESCO 1972), 911 sites around the globe have been designated as World Heritage Sites (WHSs). Although the jury is out on whether there is a link between inscription as a World Heritage Site and increased visitation (Buckley 2004; Hall and Piggin 2001), there is a rising call for the management of World Heritage Sites (Leask and Fyall 2006: xxii) due in large part to the detrimental effects of over-visitation at some sites (du Cros 2006; du Cros 2008; World Tourism Organization 2005). Exceeding the carrying capacity of certain WHSs in high-populated or high-visited countries is seen as a primary threat (ICOMOS 1993). Visitation levels and the sustainability of heritage sites have been linked, and WHS designation has been viewed as the 'catalyst to rapid tourism development via increased publicity of those destination areas' (Li et al. 2008: 308).

The formation of a potential visitor's image of a destination, or the destination image, is influenced by various information sources. One of the most simplistic and earliest definitions of destination image divides the construct into organic and induced categories, a scheme first proposed by Gunn (1972: 110). The organic component of the destination image construct is knowledge about a destination one acquires during the natural course of life, while the induced component of the destination image can be influenced or shaped by the various marketing efforts of both destinations themselves and suppliers (Baloglu and McCleary 1999; Gunn 1972). For example, Australia 'deliberately makes use of its WHS in its tourism image making' (Boyd and Timothy 2006: 56). There is a general consensus that a site's overall positive destination image plays a role in destination choice and visitation through building awareness (Milman and Pizam 1995).

Depending on the scope and intents of the publisher and how the information is accessed, information about a destination can fall into either the organic or the induced category. Information or publicity can take many forms, including print and electronic mass media, such as radio, television, movies, and, ever increasingly, websites. Some forms of publicity and promotional materials are more ephemeral, such as television advertisements, while others are more durable and long term. These include magazine and newspaper articles, archived and retrievable with the LexisNexis database, as well as popular and scholarly books, films, and maps including place names, stored in libraries and other cultural repositories around the world with references to them contained in searchable scholarly databases such as WorldCat. This durable and searchable category of documents in the public record is an indication of the level of societal reproduction.

Site sacralization and societal reproduction

According to Dean MacCannell's site sacralization theory (MacCannell 1976: 44–45), sites (i.e., destinations) become more attractive, and thus more often chosen for visitation, as they progress through five stages towards site sacralization (or destination attractiveness): naming, framing and elevation, enshrinement, mechanical reproduction, and social reproduction.

For MacCannell (1976), the naming phase, or 'marking' a site, is an important first step towards 'site sacralization.' As sightseeing objects or sites become the focal point of visitors, they are given names, which is 'the first phase of sight sacralization.' The importance of proper naming is suggested by or can be derived from anthropological and sociological studies in the universal and seminal power and role of naming and naming systems for people and places (Basso 1996; Tooker 1984) in cultures around the world. This first stage corresponds with Butler's (1980) 'exploration' stage in which tourism as such is nascent and visitor numbers are dispersed and insignificant, a stage best described as 'pre-tourism.' Framing is the 'placement of an official boundary around the object,' and elevation is 'putting an object on display' or opening it up for visitation (Jacobsen 1997; MacCannell 1976: 44). 'Elevating' a place by designating it a national park would be another example. Instead of applying Goffman's frame analytic perspective to sites (Goffman 1974), MacCannell independently took the idea of 'frame' from the fact that paintings in museums and galleries are framed before they are put on display (personal communication with MacCannell, September 17, 2009). MacCannell identified 'enshrinement' as the third stage of the evolution of attractions or destinations (Clark 2002). The boundary around the object (as a container) becomes a named entity and, eventually, a tourist attraction in its own right. An appropriate example would be the Old Faithful Geyser at Yellowstone, an attraction in itself that is surrounded by the park. This leads to an increase in tourism numbers and the attraction's reputation is enhanced (Clark 2002). Mechanical reproduction refers to the 'creation of prints, photographs, models, or effigies of the destination that are themselves valued and displayed' (MacCannell 1976: 45). For example, the creations of prominent painters, authors, and artists often have the name of the object they 'reproduced' in the title, abstract, or caption of the work. The fifth and final phase towards site sacralization, or destination attractiveness, is social reproduction, which occurs when 'groups, cities and regions begin to name themselves after famous attractions' (MacCannell 1976: 45), which is evidenced by place names on maps which can be found in the public record. For this study, these final two stages were combined and are referred to as societal reproduction.

When taking MacCannell's theory into consideration, destinations that have more WorldCat documents (i.e., those with more societal reproduction in relation to their name) should be further along in the site sacralization process and thus intrinsically more attractive sites (i.e., destinations). The number of records in durable public records such as WorldCat can be viewed as measures of public awareness of and the destination attractiveness of a site. More documents imply more overall public awareness and destination attractiveness, which should lead to increased visitation.

This study will look at whether the level of societal reproduction in relation to specific WHSs can be a reliable predictor of visitation. Because some heritage sites are under-visited, while others are endangered from unsustainable levels of over-visitation, it is prudent to investigate what factors may contribute to visitation levels in order to be able to forecast and manage future demand.

World Heritage Sites and US national parks

Many World Heritage Sites symbolize their country and its culture (Shackley 1998). Thus, it is no surprise that in the United States, the country that gave birth to the national park system with the establishment of Yellowstone in 1872, 17 of its 21 World Heritage Sites are managed by the National Park Service (NPS).[1] Those national parks designated World Heritage Sites between the late 1970s and the mid-1990s are a mixture of natural and cultural categorizations (see Table 12.1).

Because US national parks have an appropriate management program in place to preserve their outstanding universal value (UNESCO 1972), visitation statistics by park were readily available through NPS channels. Most park visitation data goes back to the time, or near it, when each park was first designated a national park. For example, visitation data for Yellowstone, established in 1872, was available from 1904. Currently, visitation data for the majority of World Heritage Sites is often unobtainable, short term or spotty, thus making system-wide analysis difficult, if not impossible. The accessibility of visitation data for US national parks was a primary reason for choosing them as WHS representatives for this study.

Methodology

Preliminary non-parametric, descriptive analysis revealed a strong positive correlation between the level of societal reproduction and visitation (see Figure 12.1) but implied no causation. Preliminary inspections of individual park statistics revealed a positive correlation between the number of documents in the public record, or societal reproduction, and visitation, though to varying degrees (see Figure 12.2). And although most had positive trends, the scale of visitation and documents differed by park. Thus, this study investigated possible causation

Table 12.1 World Heritage Sites in the United States also inscribed as national parks (by 2009), ordered by the year they became national parks

Name of property	*NP Year*	*WHS Year*	*Category*	*Location*
Yellowstone	1872	1978	Natural	Wyoming
Yosemite	1890	1984	Natural	California
Mesa Verde	1906	1978	Cultural	Colorado
Glacier (Waterton)	1910	1995	Natural	Montana
Hawaii Volcanoes	1916	1987	Natural	Hawaii
Grand Canyon	1919	1979	Natural	Arizona
Great Smoky Mountains	1926	1983	Natural	Tennessee/ North Carolina
Mammoth Cave	1926	1981	Natural	Kentucky
Carlsbad Caverns	1930	1995	Natural	New Mexico
Statue of Liberty	1933	1984	Cultural	New York
Everglades	1934	1979	Natural	Florida
Olympic	1938	1981	Natural	Washington
San Juan NHS	1949	1983	Cultural	Puerto Rico
Independence NHS	1956	1979	Cultural	Pennsylvania
Redwood	1968	1980	Natural	California
Chaco Culture	1980	1987	Cultural	New Mexico
Wrangell-St. Elias	1980	1979	Natural	Alaska

Source: Adapted from National Park Service (n.d.); UNESCO (2016).

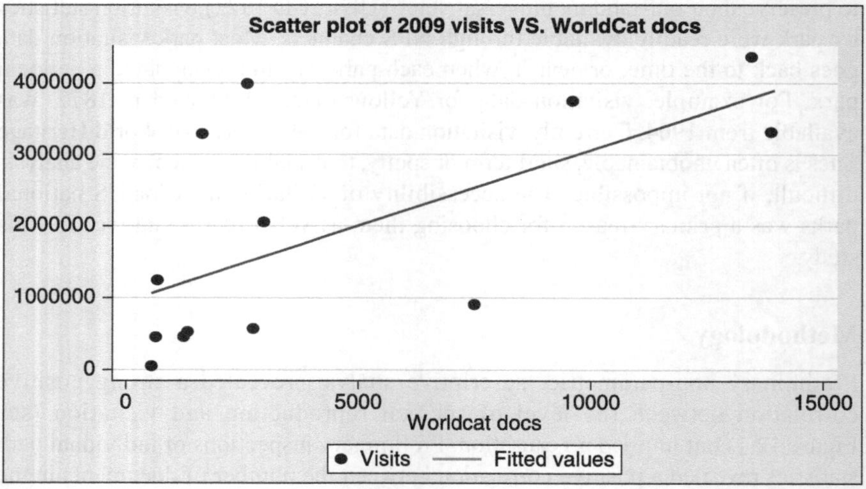

Figure 12.1 Initial scatter plot of 2009 data for all 13 national parks that are also WHSs suggested a strong positive correlation between the count of documents retrieved from WorldCat and visitation levels.

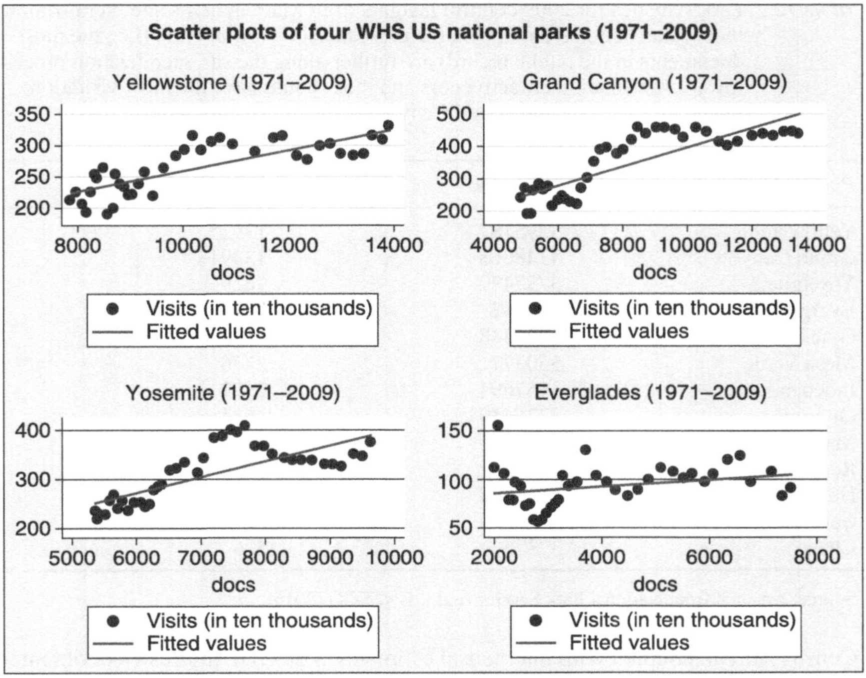

Figure 12.2 Scatter plots of the count of documents in the public record retrieved from WorldCat from 1971 to 2009 vs. visitation for the four national parks with the most retrieved documents show a strong positive correlation.

using regression analysis with visitation and document data for 39 years, the time period for which visitation statistics were available for all of the parks in the study.

The two key variables in this research were a measure of societal reproduction and 1971–2009 visitation levels for a sample of US national parks that are also World Heritage Sites (WHSs). Societal reproduction was operationalized as the count of objects retrieved from WorldCat, an online catalog of over 49 million objects in repositories worldwide, where a park's name was entered as a keyword. In order to capture all or most records in the database, the year 1700 was established as the base. For example, to extract the number of records in WorldCat for Yellowstone for the year 1971, the date range was constrained from 1700–1971, the keyword was Yellowstone, and no limits were placed on language, type of library, or material. This method was repeated for all 39 years. The number of 'All' retrieved documents was recorded in an Excel spreadsheet. Retrieved objects include books, historical photographs, archival documents, journals, dissertations, maps (which include place names), audio-visual materials such as documentaries, videos, and films, and thesis and other manuscripts, recorded in the public record.

The population of the United States and Gross Domestic Product (GDP) per capita served as controls. The population data consists of figures from the US

Table 12.2 Park statistics for 2009 confirm insights from MacCannell's site sacralization theory that national parks with the most societal reproduction (i.e., the most documents in the public record) are further along the site sacralization process towards destination attractiveness and thus should have the most visitation; iconic Yellowstone and Grand Canyon are the most attractive and most visited sites

Park	*Visitation*	*WorldCat docs*
Yellowstone	3295187	13925
Grand Canyon	4348068	13491
Yosemite	3737472	9629
Everglades	900882	7515
Glacier	2031348	2975
Mesa Verde	550377	2776
Independence	3967694	2629
Olympic	3276459	1684
Mammoth	503856	1389
Redwood	444426	1269
Hawaii Volcanoes	1233105	735
Carlsbad	432639	677
Chaco	37376	576

Source: Adapted from National Park Service (n.d.); UNESCO (2016).

Census Bureau coupled with intercensal estimates. The GDP figures were obtained from World Bank's World Development Indicators database and the *CIA Factbook*. Although visitation and public record document data was collected for all 17 WHS US national parks, a case study analysis was completed with a subset of the four sites with the most retrieved docs: Yellowstone, Grand Canyon, Yosemite, and Everglades (see Table 12.2). This was prudent because, if applying MacCannell's theory, the parks with the most related documents in the public record should be the most 'sacralized,' or more advanced along the continuum towards site sacralization and thus destination attractiveness. This subset of parks also contained some of the most iconic tourism destinations in the United States, and ones for which place names are synonymous with the site. The final dataset consisted of 156 cases ($n = 156$), which is over the threshold of 30 recommended cases for regression analysis. Because the variables were not normally distributed and in different units, they were inspected for suitable transformations to obtain normality, which revealed a log transformation would be appropriate. The resulting double log model produced coefficients that represent elasticity or percent change, which provide a straightforward interpretation. Regression and graph creation was conducted with Stata 11.

Results

The double log model comprising the log of visitation (lnvisits) and the log of WorldCat docs (lndocs) produced an R^2 of 60.85, which means that approximately 61 percent of the variation in visitation can be explained by the count of documents

Table 12.3 Impact of count of retrieved documents in the public record where a park's name is entered into WorldCat on visitation from 1971 to 2009 at four iconic US national parks that are also World Heritage Sites

					Number of obs = 156 *F(1, 154) = 239.38* *Prob > F = 0.0000* *R-squared = 0.6085* *Adj R-squared = 0.6060* *Root MSE = 0.35782*	
lnvisits	*Coef.*	*Std. Err.*	*t*	*P>t*	*[95% Conf.*	*Interval]*
lndocs	**0.9626696**	**0.0622203**	**15.47**	**0.00**	**0.8397543**	**1.085585**
_cons	**6.114271**	**0.5500646**	**11.12**	**0.00**	**5.027625**	**7.200917**

(see Table 12.3) or the level of societal reproduction, which plays a role in destination choice and attractiveness. The adjusted R^2 is very close to the observed value, which indicates a good cross-validity of the model. This is a very strong result and thus supports the hypothesis that the count of retrieved documents from WorldCat may be a good predictor of visitation. The coefficient for 'lndocs' is 0.9626696, meaning that for every unit increase in 'lndocs' we would expect a 0.96 unit increase in 'visits.' The probability value or *p*-value below 0.05 for the 'lndocs' variable means that the test is statistically significant. The model assumption of constant variance for residuals was checked with a Breusch-Pagan/Cook-Weisberg test for heteroskedasticity. According to the results, the small chi-square ($\chi^2 = 3.13$, Prob > $\chi^2 = 0.0769$), indicates no heteroskedasticity was present. The results are significant at the $p = 0.05$ level.

The results of stepwise regression with the additional variables of the log of US population (lnuspop) and the log of GDP (lngdp) revealed that only 6 percent of the variation in visitation was explained by rising US population and GDP, reconfirming that the majority of the variation in visitation could be attributed to the influence of the number of retrieved WorldCat documents (see Table 12.4). In fact, a negative coefficient for population (lnuspop) suggests that visitation to US parks has not been keeping pace with population growth in the United States, which could be the base of another study. The influence of GDP was positive, but minimal in effect. The coefficient for 'lndocs' is 1.176903, meaning that for every unit increase in 'lndocs' we would expect a 1.18 unit increase in 'visits.' For example, for every 1 percent increase in documents in the public record in a given year, one can expect a 1.18 increase in visitation. The model assumption of constant variance for residuals was checked with a Breusch-Pagan/Cook-Weisberg test for heteroskedasticity. According to the results, the small chi-square ($\chi^2 = 0.50$, Prob > $\chi^2 = 0.4817$) indicates no heteroskedasticity was present. The results are significant at the $p = 0.05$ level.

Applying the results of the regression analysis, one finds that if a World Heritage Site manager or a gateway community was interested in increasing

Table 12.4 Impact of count of retrieved documents in the public record where a park's name is entered into WorldCat on visitation from 1971 to 2009 at four iconic US national parks that are also World Heritage Sites with controls

					Number of obs = 156	
					F(1, 154) = 102.07	
					Prob > F = 0.0000	
					R-squared = 0.6683	
					Adj R-squared = 0.6617	
					Root MSE = 0.33154	
lnvisits	*Coef.*	*Std. Err.*	*t*	*P>t*	*[95% Conf.*	*Interval]*
lndocs	**1.176903**	**0.0708659**	**16.61**	**0.000**	**1.036894**	**1.316913**
lnuspop	**–2.24333**	**1.008877**	**–2.22**	**0.028**	**–4.236562**	**–0.2500986**
lnusgdp	**0.1529605**	**0.1837776**	**0.83**	**0.407**	**–0.2101279**	**0.5160488**
_cons	**46.10208**	**17.65861**	**2.61**	**0.010**	**11.21408**	**80.99008**

visitation by raising public awareness of the site, a relatively modest 1 percent increase in recorded and retrievable documents in the durable public record would lead to approximately a 1 percent increase in visitation. For example, a site with a low level of documents and comparably low visitation, such as Chaco Culture in 2009 (docs = 576, visitation = 37,376), could expect to achieve a predicted 1.18 percent increase in visitation, or an additional 441 visits, if only six documents were added to the public record annually. Thus, a managing entity could support ways to increase the creation and flow of site-related documents into the durable public record. Conversely, if a site manager wanted to stem the tide of visitation in order not to exceed a site's carrying capacity, one could look for strategies to limit or slow the addition of site-specific documents entering the durable public record.

Discussion

This study sheds light on and brings attention to several issues related to the management and marketing of WHSs. The level of societal reproduction in relation to a site can be viewed as a reliable predictor of visitation, as well as a proxy for destination attractiveness. In a practical sense, an increase in documents in the durable public record could be an indicator of increased visitation to come, thus giving managers time to prepare and respond. But is it prudent or possible to control the creation and flow of documents in the hope of controlling visitation by manipulating a site's public awareness level or its destination attractiveness? Could limiting the sponsorship of journalists or authors who want to research, write about, and ultimately promote World Heritage Sites, acts that would conceivably lead to the addition of documents to the durable public record (which would increase attractiveness because consumers would become more aware of the destination and thereby lead to increased visitation), result in less visitation? Would limiting advertising, another kind of information dispersal, slow demand? What is the right strategy and balance of strategies for a particular site?

Some events that contribute documents to the durable public record that might influence future visitation are unforeseeable and uncontrollable. A recent example is the production and publishing of Ken Burns' documentary and accompanying coffee table book on the founding and importance of US national parks (Burns and Duncan 2009). The Public Affairs Officer at Grand Teton National Park, a sister park to Yellowstone, voiced the desire to detect increased visitation by conducting a visitor survey in the summer of 2010, although it was not feasible because the park had just co-sponsored a survey in 2009 (personal communication with Jackie Skaggs, May 14, 2010). If such a survey had taken place, Skaggs would have included a question about whether Ken Burns' new book and film had inspired visitation.

At the UNESCO conference in Quebec (June 2–4, 2010), restricting access to the use of World Heritage Sites through implementing some sort of copyright protection for the owners of World Heritage Sites was proposed (Cameron 2010), an idea that dovetails well with the findings of this research. This could be implemented to protect both place names and images associated with WHSs, including sites contested by indigenous groups (Brown 2003: 16), such as Australia's Uluru. For a very long time, photographers have been encouraged, and sometimes required by publications, to get signed releases from their human subjects before they can sell, print, or post their images. It is not a huge leap to take lessons from the publishing 'release' model to implement some sort of copyright or signed release for World Heritage Sites. If the implications of this study hold true, it is probable that over time such measures would help reduce the flow of documents into the durable public record or, in other words, moderate the level of societal reproduction in relation to specific sites (i.e., including films and books with WHS settings and place names), ultimately reducing visitation. Other strategies could also be explored.

Limitations and suggestions

Lack of consistent and long-term visitation data for World Heritage Sites in many countries may make this study hard to duplicate in other settings. It may be likely that there is a symbiotic relationship between publicity about a specific WHS and visitation, but this kind of effect would be very hard to capture empirically. Also, all of the sites in this study were national parks well *before* they were designated World Heritage Sites. For example, Yellowstone was an iconic destination for more than 100 years before WHS designation. Thus, these results may not apply uniformly to sites that had relatively little visitation prior to WHS designation.

This study could be expanded or changed to investigate the relationship between visitation and the count of magazine and newspaper articles stored in and retrievable from the LexisNexis database. WorldCat was manageable, scholarly, and finite, but this study could also be expanded or refocused to examine the influence of Web 2.0 public records in relation to visitation at WHSs, such as those retrieved through Google's search engine or any number of other online searchable databases and resources. For example, UNESCO's World Heritage

Center has entered into a partnership with TripAdvisor (owned by Expedia.com) 'to raise awareness and protect World Heritage Sites' (UNESCO 2009). Expedia.com clients can book heritage trips through their website and have the opportunity to post online site evaluations at the sister site TripAdvisor.com. However, as this study implies, raising public awareness may lead to more visitation, which does not always go hand-in-hand with protection or sustainability. Thus, it would be interesting to investigate the effect of this partnership.

Note

1 World Heritage Site status is not capitalized on in the United States as in other countries, so much so that most park tourists do not know they are visiting a World Heritage Site (Timothy and Boyd 2006: 246–247).

References

Baloglu, S. and McCleary, K. (1999) 'A Model of Destination Image Formation', *Annals of Tourism Research*, 26 (4): 868–897.

Basso, K. (1996) *Wisdom Sits in Places: Landscape and Language among the Western Apache,* Albuquerque: University of New Mexico Press.

Boyd, S. and Timothy, D. (2006) 'Marketing Issues and World Heritage Sites', in A. Leask and A. Fyall (eds.), *Managing World Heritage Sites*, Burlington, MA: Butterworth-Heinemann.

Brown, M.F. (2003) *Who Owns Native Culture?* Cambridge, MA: Harvard University Press.

Buckley, R. (2004) 'The Effects of World Heritage Listing on Tourism to Australian National Parks', *Journal of Sustainable Tourism*, 12 (1): 70–84.

Burns, K. and Duncan, D. (2009) *The National Parks, America's Best Idea: an illustrated history.* New York: Alfred A. Knopf.

Butler, R.W. (1980) 'The Concept of a Tourist Area Cycle of Evolution: Implications for Management of Resources', *Canadian Geographer*, 24 (1): 5–12.

Cameron, C. (2010) 'World Heritage Tourism, Right or Privilege?' Paper presented at the World Heritage and Tourism: Managing for the Global and Local Conference, Quebec City, Canada, June 2–4, 2010.

Clark, I.D. (2002) 'The Ebb and Flow of Tourism at Lal Lal Falls, Victoria: A Tourism History of a Sacred Aboriginal Site', *Australian Aboriginal Studies*, Fall: 45–53.

du Cros, H. (2006) 'Managing Visitor Impacts at Lijiang, China', in A. Leask and A. Fyall (eds.) *Managing World Heritage Sites*, Burlington, MA: Butterworth-Heinemann.

du Cros, H. (2008) 'Too Much of a Good Thing? Visitor Congestion Management Issues for Popular World Heritage Tourist Attractions', *Journal of Heritage Tourism*, 2 (3): 225–238.

Goffman, E. (1974) *Frame Analysis: An Essay on the Organization of Experience*, Cambridge, MA: Harvard University Press.

Gunn, C. (1972) *Vacationscape: Designing Tourist Regions*, Austin: University of Texas at Austin.

Hall, C.M. and Piggin, R. (2001) 'Tourism and World Heritage in OECD Countries', *Tourism Recreation Research*, 26 (1): 103–105.

ICOMOS (1993) *Cultural Tourism: Tourism at World Heritage Cultural Sites. The Site Manager's Handbook* (2nd ed.), Madrid: World Tourism Organization.

Jacobsen, J. (1997) 'The Making of an Attraction: The Case of North Cape', *Annals of Tourism Research*, 24 (2): 341–356.
Leask, A. and Fyall, A. (2006) *Managing World Heritage Sites*, Burlington, MA: Butterworth-Heinemann.
Li, M., Wu, B. and Cai, L. (2008) 'Tourism Development of World Heritage Sites in China: A Geographic Perspective', *Tourism Management*, 29 (2): 308–319.
MacCannell, D. (1976) *The Tourist: A New Theory of the Leisure Class*, New York: Schocken Books.
Milman, A. and Pizam, A. (1995) 'The Role of Awareness and Familiarity with a Destination: The Central Florida Case', *Journal of Travel Research*, 33 (3): 21–28.
National Park Service (n.d.) Welcome to Visitor Use Statistics. US Department of the Interior. Online, https://irma.nps.gov/Stats/ (accessed May 9, 2016).
Shackley, M. (1998) *Visitor Management: Case Studies from World Heritage Sites*, Oxford: Butterworth-Heinemann.
Timothy, D. and Boyd, S. (2006) 'World Heritage Sites in the Americas', in A. Leask and A. Fyall (eds.), *Managing World Heritage Sites*, Burlington, MA: Butterworth-Heinemann.
Tooker, E. (1984) 'Naming Systems: 1980 Proceedings of the American Ethnological Society.' Paper presented at the American Ethnological Society Central States Anthropological Society Association for the Anthropological Study of Play, Washington, DC.
UNESCO (1972) Convention Concerning the Protection of World Cultural and Natural Heritage. Online, http://whc.UNESCO.org/?cid=175 (accessed August 30, 2010).
UNESCO (2009) 'UNESCO and Online Travel Giant TripAdvisor Launch World Heritage Partnership', November 23, 2009. Online, www.UNESCO.org/new/en/media-services/single-view/news/UNESCO_and_online_travel_giant_tripadvisor_launch_world_heritage_partnership/ (accessed August 17, 2010).
UNESCO (2016) United States of America: Properties Inscribed on the World Heritage. Online, http://whc.unesco.org/en/statesparties/us (accessed May 9, 2016).
World Tourism Organization (2005) *Tourism Congestion Management at Natural and Cultural Sites*. Madrid: World Tourism Organization.

13 Looking back towards the future

Historical analysis of Machu Picchu planning documents as a key to site conservation

Evan R. Ward

El turismo es como el fuego: puede hacer hervir su olla o bien incendiar su casa.[1]
(Lovón Zavala 1982: 1)

In 2007, publication of Alonzo C. Addison's *Disappearing World* dramatically illustrated the human and natural threats posed to World Heritage Sites (WHSs) around the world. In a number of cases Addison cited tourism-related factors that threatened WHS archaeological and architectural ensembles (Addison 2007). While technical specialists will be called upon to implement more sustainable practices at these sites, historians offer important insight regarding the rehabilitation of such sites.

Although the United Nations for Educational, Scientific and Cultural Organization (UNESCO) nominated WHSs only after 1972, many of them benefited from UNESCO programs initiated a decade earlier. In the mid-1960s, in an effort to improve the economic standing of emerging nations, UNESCO sponsored a series of technical missions to advise national governments on how they could simultaneously preserve cultural patrimony and stimulate cultural tourism. These consultations coincided with the United Nations' designation of 1967 as the 'Year of International Tourism' (UNESCO 1966: 5–10). The consultants selected to carry out these missions represented diverse backgrounds, but all were charged with assisting emerging nations in preparing cultural patrimony for international tourism. Reviewing the suggested courses of action, as well as subsequent actions taken by the host governments, may better help practitioners to understand the value systems upon which such sites were preserved and presented to the public. More importantly, these documents reveal the 'planner's gaze,' a critical component of the backstage production of tourist attractions (adapted from Urry 2009; MacCannell 1999: 45). In contrast to tourists, who view the touristic stage created for mass consumption, planners bring their own set of values to the production of tourism. This 'planner's gaze' is corroborated in the fieldwork of A. Frischlander's 1968 consultation at Machu Picchu, where he observed that 'the arrival at an archeological site is of high importance as the arrival should be for the visitor the culminating moment of the visit . . . The ruins should appear gradually before the eyes of the visitor, marveling at its grandiose aspects' (Frischlander 1968: 6). Thus, planners play a significant role in shaping the ways in which tourists look upon a given tourist site. Nevertheless, an individual planner's 'gaze' is usually contested by the sponsoring government, other consultants, and commercial influences that shape the reproduction of how such sites are staged and consumed. Thus, while the planner's vision of how a site should be designed and viewed is generally contested, planners play a prominent role in shaping a project's orientation, if not specific ways in which tourists 'see' an attraction (MacCannell 1999: 91–107).

Planning for tourists at Machu Picchu

The planning processes for the Machu Picchu sanctuary are well documented and lend themselves to historical analysis. Chronologically, state planning for tourism development in Peru began in the mid-1960s. While previous private-sector initiatives placed Machu Picchu on the global tourist map prior to this time, state responsibility for leisure planning further vindicated the site's status as a world-class tourist attraction. To provide some context, it is important to recall that the mid-1960s were the heyday of modernization theory and its application to tourism development. Hopes ran high in developing nations that tourism could make a substantial contribution to economic development by reducing unemployment and increasing exchange reserves (Aguilar et al. 1992; Defert and Pichot 1988: 2–5). In 1965, the Peruvian government asked the Organization of American States (OAS) to provide recommendations for tourism development throughout the country. The OAS gave the assignment to Harry G. Clement, a long-time tourism industry consultant most famous for conducting a 17-nation survey of tourism in the Asian Pacific for the Pacific Area Tourism Association between 1958 and 1961 (Clement 1961). Clement believed that successful tourism policies could easily be transferred from developed nations to developing countries with little difference in outcomes.

Needless to say, Clement's recommendations for Peruvian tourism development mirrored those from his Asian study. Namely, the private and public sectors needed to cultivate tourism attractions on the Peruvian periphery, as well as build new hotels to accommodate international tourists. Clement quickly identified the Cuzco area as the prime tourist destination in Peru outside of Lima:

> [It's] really Machu Picchu that is the principal lure here. To the extent that Peru has a single attraction that is known throughout the world, it's Machu Picchu. And in terms of promotion, Machu Picchu is a salable item, particularly when sold in tandem with Cuzco.
>
> (Clement 1965: 29–30)

Clement's disturbing commodification of Machu Picchu suggests that a long-term private-sector campaign had established the site's success prior to state planning there. Therefore, to properly understand the state-directed planning of Machu

Picchu, one must backtrack 50 years to understand its commercial framing. Once the record of Machu Picchu's commercial development is better understood, it becomes much easier to comprehend why politicians and planners consistently chose a less sustainable path to development during the twentieth century. Conversely, the same analysis brings to light discarded recommendations for the site that could improve its sustainability.

Hiram Bingham: prelude to tourism

The commodification of Machu Picchu began with the activities of Hiram Bingham in Peru (Bingham 2003; Heaney 2010). Bingham's close connections with W.R. Grace & Company during his travels in South America are at the root of this process. What is often not mentioned about Bingham is that he was a historian, not an archaeologist, and originally traveled to South America in 1909 to attend a conference and follow the path of revolutionary heroes across the Americas. Following that excursion, he noted of his travels in Bolivia:

> Fortunately we were not obliged to experiment upon the proverbial Bolivian hospitality, but were met at the station by kind friends, representatives of W.R. Grace and Company, who did everything in their power to add to the debt of gratitude which I had owed their house ever since I started my journey. Comfortable quarters were found for us in the Sucursal, a huge, modern, three-story building intended for a convent, but now used as the annex of the leading hotel.
>
> (Bingham 1911a: 231)

Sounding more like a tourist than an explorer, Bingham continued to cultivate his relationship with W.R. Grace during his subsequent expeditions to Peru. Prior to his largely self-funded trip in 1911, he importuned a W.R. Grace official for financial support for the expedition, but later expressed even greater gratitude to W.R. Grace officials for extricating him from the reach of Peruvian authorities in carrying out his excavations (Bingham 1911b; Bingham 1911c)

A review of the two boxes of Bingham's correspondence with W.R. Grace officials prior to, during, and immediately after his 1911 expedition erases the possibility that the shipping magnate viewed Bingham's expedition as an avenue for promoting tourism in Peru. However, company advertising, published in the August 1911 edition of *National Geographic* (a month after Bingham's first visit to Machu Picchu), clearly established W.R. Grace's interest in promoting tourism to Peru. This curious advertisement, unlike any other later published by the company in *National Geographic*, provides an indirect connection between Bingham's reliance on the company during his initial trip to Machu Picchu and W.R. Grace's interests in developing Peruvian tourism. The ad imbued the land of Pizarro with a sense of nostalgia for the Spanish conquistadors who vanquished the Incan empire with sword and cross. 'Such a land,' the advertisement suggested, "cannot but offer irresistible attractions to the modern tourist' (W.R. Grace & Company 1911: 787).

Ironically, perhaps Bingham's greatest contribution to the popularity of Machu Picchu was his amateur status as an archaeologist. In the same vague fashion in which W.R. Grace and writers portrayed the vanquished lands of the Inca, Bingham's inconclusive claims about the origins of Machu Picchu left the door open for tourists to draw their own conclusions about the meaning of the site. Much like the 'seduction of difference,' this 'seduction of ambiguity' served as a powerful attraction to tourists who might believe that 'seeing' the site for themselves would resolve any differences between the claims of competing markers that have been attached to Machu Picchu over time. Ultimately, however, W.R. Grace bears significant responsibility for promoting international tourism to the site. As such, W.R. Grace's saturation of American print media with images of Machu Picchu should be added to Geoffrey Shullenburger's assertion that 'Bingham's international publicizing of Machu Picchu allowed for a seamless and near immediate transition into the tourist circuit' (Shullenberger 2008: 320).

W.R. Grace's promotion of Peru in general and Machu Picchu in particular ebbed until the post-World War II era. Thereafter, W.R. Grace's airline, Panagra (a partnership with Pan Am), moved the commercial process of site sacralization towards completion through intense advertising efforts (MacCannell 1999). In one ad, for example, Panagra advertisers placed Machu Picchu at the center of a collage of familiar tourist icons accompanied by the statement: 'You're not a world traveler until you can name these eight places!' The ad's author presumed that Machu Picchu had already gone through the process of site sacralization and was now ascending the list of 'must-see' destinations (Panagra 1959: 315). Like Bingham, Panagra executives, in the words of Shullenberger,

> [appeared] to have occasioned a rift in the construction of an archaeological patrimony in which a single, largely decontextualized object [the Machu Picchu sanctuary], having been converted into an endlessly reproducible image, [achieved] an astonishing and unparalleled prominence in the global public sphere.
>
> (Shullenberger 2008: 320)

Thus, beginning in the late 1940s, Panagra blitzed *National Geographic* and prominent newspapers with more than 20 attractive visual and textual promotions of Machu Picchu or Cuzco. By 1962, travel writer Robert Kane observed: 'There are probably more postcards written home exclaiming over the wonders of Cuzco and Machu Picchu than any other tourist destinations in South America, and with good reason' (Kane 1962: 340).

Panagra's efforts to promote travel to Machu Picchu also stepped across the fuzzy line separating mass media advertisements and public relations. In November 1953, Museum of Modern Art Director Rene d'Harnoncourt invited Panagra and Grace Lines (the shipping arm of W.R. Grace & Company) to provide transportation of exhibit pieces for its upcoming 'Ancient Arts of the Andes' exhibition from Europe and South America to New York City. In exchange, d'Harnoncourt pledged: 'Credit would, of course, be given to Panagra and Grace

Line for their assistance in both the publication, the exhibition, and general publicity' (d'Harnoncourt 1953). In February of 1954, Panagra Public Relations Director Louis J. Garcia forwarded three advertisements featuring Panagra on its WQXR radio program to d'Harnoncourt. The first two focused specifically on the exhibition itself, but the third made a strong connection between the airline and Machu Picchu. Visitors to the museum might not only envision the storied city in the historical imaginary, but also consider how 'a Panagra plane could easily carry these stones thousands of miles at 300 miles an hour!' Subsequently, a visit to the exhibit in New York might encourage the museum-goer to take a flight to Peru on Panagra for purposes of visiting Machu Picchu in person (Garcia 1954).

By the late 1960s, Panagra and Pan Am sold their South American air routes to Braniff Airlines, a transition which did nothing to slow the hyping of Machu Picchu as 'the next place' (Braniff International 1968).

Alternatives to mass international tourism

At roughly the same time that Panagra's promotional activities were winding down, Harry G. Clement made his recommendations to improve international tourism in Peru. Clement's commercial representation of Machu Picchu and its placement near the summit of Peruvian tourism were likely influenced by Panagra's promotion of Machu Picchu. Aside from Clement's published consultations of tourism development throughout the world, few clues remain of his past that might give context to his statements. However, a contemporary of Clement, Basil Atkinson, a high-ranking Australian tourism official in the 1960s, noted that one of his friends, John Payne, a former Pan Am vice president, was a 'firm [advocate] for the commissioning and then the implementing of major recommendations' in Clement's strategies in Asia (Atkinson 2009). The focus on long distance tourism from the United States, as well as the construction of luxury hotels in Peru, also suggest overlap in the objectives of Panagra and Clement there.

Clement, however, was not the only source from which the Peruvian government would seek advice for developing tourism. Thus, the 'planner's gaze' could as easily be contested by opposing consultants as by the state. At approximately the same time that the Fernando Belaúnde Terry administration commissioned the OAS study, Peruvian officials also sought out the advice of UNESCO, which had recently taken an interest in promoting the conservation of cultural patrimony as a tool for economic development (Villesa Lescano 2007: 46–90). Unlike the OAS consultation, the UNESCO studies involved a handful of specialists from different disciplines. As a result, UNESCO consultants provided the Peruvian government with a broader range of possibilities for framing tourism development. In essence, these consultants offered competing conceptualizations of what type of tourism would best benefit Peruvians. For example, French consultant Pierre Defert, pioneer of tourism geography, conducted an economic evaluation of 'The Development of Peru's Historic Sites and Monuments.' While he agreed with Clement on various points, most importantly the need for more hotels throughout the country, he also insisted that the Peruvian government needed to make sure that the infrastructure and archaeological work needed at Machu Picchu be completed before it was promoted around the world (Defert 1967: 3–7). Unfortunately, Defert did not take into account that commercial interests like Panagra had initiated the promotion of Machu Picchu as early as the 1940s. For the next half-century (after 1965), private and public entities would attempt to meet the demand of tourists there, falling short more often than not.

British urban planner Graeme Shankland recommended that the Peruvian government promote a wide number of archaeological sites around Machu Picchu and Cuzco, rather than narrowly focusing on Machu Picchu. In essence, he believed that Machu Picchu could be the starting point for an equally rewarding tourist circuit linking Cuzco, Machu Picchu, and other archeological and natural sites in the region (Shankland 1969: 18–19). Similarly, F. Engel's study 'Protection and Exploitation of Historical Monuments' explored alternative forms of tourism in the Cuzco region. While focusing on international tourists would indeed be profitable, he noted, attracting tourists closer to home offered greater long-term sustainability. He observed:

> In general, the idea of an invasion by tourists with limited means . . . was considered preferable to that of a handful of rich tourists living clannishly together in a luxury hotel: tourists of this kind are generally criticized for their lack of contact with the population at large.
>
> (Engel 1967: 8)

In order to accomplish this, the Peruvian government would need to improve roads and create 'staging posts [hotels and restaurants] organized along acceptable lines and kept within a price range suiting the pockets of tourists with modest means' (Engel 1967: 8). Ultimately, Shankland and Engel offered solutions that featured a more diverse set of attractions for tourists, as well as a more diverse tourist mix.

If Harry G. Clement's report represented an initial assessment of tourism strategies in Peru, and the UNESCO reports served as a collective anti-thesis to Clement's report, then the Peruvian government's publication *La actividad turística y el desarrollo económico y social del Perú: un estudio de diagnóstico* could be considered the synthesis for Peruvian tourism in the early 1970s (COPESCO c.1973). In this document, the Peruvian government repudiated Clement's economic model, based on a 'multiplier effect,' yet accepted Cuzco/Machu Picchu as the geographic focus of Peruvian tourism (COPESCO c.1973: 35). Adapting tourism development to Peruvian realities, the socialist Velasco Alvarado government accepted UNESCO's recommendation that tourism development could serve as a spur to regional development, with tourism development transforming Peru's rural economy (COPESCO c.1973: 62). What should not be lost in the analysis of Peruvian tourism development is the fact that Peru's successive administrations effectively followed the same model for tourism development in Peru (El Congreso del Peru 2009).

Several sources corroborate the claim that both the right and the left allowed commercial interests to promote Machu Picchu as a mono-crop tourism product. While it is true that Peru cultivated other tourist sites, one Peruvian tourism official estimated, in 1995, that 98 percent of international tourism arrivals to Peru visited Machu Picchu (Deforges 2000: 178). Urban planner Giorgio Lombardi also provided proof of this commercial emphasis while on assignment for UNESCO in 1977. Lombardi observed that tourism in the region was still an 'international [affair], rigidly organized with respect to . . . an international structure of agencies, flight plans and hotels, which determine the chain of touristic attractions' (Lombardi 1977: 8). Lombardi encouraged the conversion of existing buildings in Cuzco into small hotels in order to preserve the city's viability as a place to live (Lombardi 1977: 11–12). Echoing Frederick Engel's recommendations, Lombardi also encouraged Peruvian officials to attract more South American and youth tourists – as opposed to focusing primarily on international tourists.

Local impacts of international tourism development

Perhaps the most conclusive evidence that Machu Picchu's predominantly commercial framing by planners, politicians, and corporate interests endured for nearly a century emanated from local voices in Cuzco. During the early twentieth century, the local *Indigenismo* movement attached its own values and meaning to the legacy of the Inca. As Alexandra Arrellano has ably articulated, *Indigenismo* simultaneously served as a model for economic development, political empowerment, and the shaping of cultural identity for *cusqueños* (this movement would play a direct role in the development of cultural tourism with events such as the staging of Inti Raymi in Cuzco) (Arrellano 2008: 39–56). By the late twentieth century, local activists and scholars – who served in a sense as the heirs of the spirit of *Indigenismo* – provided the most persuasive critiques of the ways in which tourism had failed residents in the Cuzco region. In the early 1980s, researchers Gerardo Lovón Zavala and Rocío Moscoso Blanco undertook a study measuring tourism's contributions to Cuzco's economy between 1950 and 1980. The authors observed that North Americans constituted approximately 30 percent of the tourist arrivals, with Western Europeans and Japanese tourists also accounting for many visitors to the region. Despite the construction of hotel rooms and the emergence of new tourism enterprises, Lovón Zavala and Moscoso Blanco concluded that tourism had not remedied the region's unemployment problems (Lovón Zavala and Moscoso Blanco 1981: 3–5). In 1982, Lovón Zavala further reflected on the legacy of tourism in Cuzco. He questioned the viability of international tourism on the Peruvian periphery, particularly since the groups least benefited by the growth of international tourism – namely, subsistence farmers – were the very people that tourism was supposed to help, but had not. He contended that the employment generated through tourism corresponded to government-sponsored jobs created by COPESCO. The promised private sector synergies had not developed, leaving the sector largely in the hands of large corporations. Lovón

Zavala summed up his 1982 study of tourism in the Cuzco region with this emphatic statement: 'Whoever has the conviction that tourism is the only alternative for Cusco in the year 2000, has a great challenge: to move from the apologetic discourse favorable to tourism and contrast their optimistic ideas and alternatives with the cold reality' (Lovón Zavala 1982: 29). Ultimately, these ineffectual tourism policies revealed the inability of national policies to generate economic development through tourism.

Conclusion

In hindsight, it is not surprising that Harry Clement saw the potential for mass tourism at Machu Picchu. Clement walked with tourism industry specialists who were concerned with the development of luxury hotels, as well as executives from the primary air carrier from the United States, Pan American Airlines. In sum, Clement's consultation for the Peruvian government suggested that international tourism in Peru, centered on Lima and side trips to Machu Picchu, offered the most immediate infusion of currency into the Peruvian economy. Nevertheless, the discursive conflict between Clement's 'gaze' and UNESCO consultants offers an excellent case study of how the planning process reveals a multiplicity of motives that surround the development of a WHS. While many of these consultations have been forgotten, they offer insight into charting the path forward as WHSs begin to face greater challenges related to sustainability.

In the case of Machu Picchu, some have argued that its recent decline is largely the product of neoliberal administrations during the 1990s and early twenty-first century (Deforges 2000). While the statements and policies of Peruvian figureheads, such as Alan Garcia, have certainly contributed to the site's inundation with visitors, this chapter argues that Machu Picchu's path to non-sustainability was set in motion during its commercial exploitation following World War II, and then confirmed during the early 1960s and 1970s, as successive governments made choices to focus tourism in the Cuzco region. Over time this singular focus has eschewed the broader initiative of a balanced tourism development program featuring a wider ensemble of archaeological sites throughout the nation.

Historical analysis offers more than a lens for viewing where planning went awry. Reviewing the initial planning stage for all WHSs offers current planners, politicians, and practitioners an opportunity to assess the evolving nature of each site. Even when a dominant trend has set the stage for tourism development around a given WHS, a periodic review of the planning phase may reveal alternatives from the perspective of the 'planner's gaze' that could have utility. This is certainly the case with Machu Picchu. The calls of its consultants for promotion of a wider variety of tourism poles, as well as closer links between tourism development and regional economic development, may suggest a more sustainable future in the Andes – if only the political will to carry out such steps can be mustered (COPESCO c.2007; Mould de Pease 2007). This is nowhere more evident than in F. Engel's suggestions for the region. Engel's preference for increasing South American tourism in the Cuzco region by constructing roads to offer maximum access to multiple tourism destinations was also a call to limit international tourism. For the past several decades, planners have debated the desirability of a new international airport in Cuzco that could accommodate long-range jets, thus facilitating the arrival of even more international tourists (*Andina* 2010). Engel may have stated his concern differently, but the question of diversifying the mix of tourists persists. Similarly, three decades have passed since Giorgio Lombardi issued his condemnation of the lack of urban planning in Cuzco. However, his recommendations, including safeguarding the viability of the colonial city, might still hold promise if officials consider the long-term benefits of promoting local economic growth over catering to international hotel chains and tour operators (*Andina* 2008).

In closing, the author would like to recommend that planning documents be consulted in looking for solutions to contemporary problems at WHSs throughout the world. The most effective review of development and planning tools would involve site managers, local stakeholders, and politicians who impact tourism development decisions. Perhaps reviews could take place six months to a year after each cycle of elections in host nations, allowing time for appointees to take office, and for elected office holders to familiarize themselves with their jurisdictional responsibilities. A review of relevant UNESCO planning materials, as well as a discussion of how such materials could be relevant to future actions at the site, would be appropriate. While cynics may discount the value of these steps, I am certain that a review of history (provided there is political will to implement any recommendations) can make a difference, from time to time, in charting new objectives for effective site management.

Note

1. Tourism is like fire: it can boil the pot or burn your house.

References

Addison, A.C. (2007) *Disappearing World: 101 of the Earth's Most Extraordinary and Endangered Places*, London: HarperCollins Publishers.

Aguilar, V., Leonith, H., and Milla, C. (1992) *Turismo y desarrollo: posibilidades en la región Inka*, Cusco: Centro de estudios regionales andinos 'Bartolome de las Casas,' pp. 8–9, 31.

Andina (December 3, 2008) 'J.W. Marriott to Build $20 Million Five-Star Hotel in Cusco.' Online, www.andina.com.pe (accessed January 10, 2012).

—— (July 16, 2010) 'Nuevo aeropuerto de Chinchero impulsará vuelos internacionales directos a Cusco.' Online, www.andina.com.pe (accessed January 10, 2012).

Arrellano, A. (2008) 'The Inca Heritage Revival: Indigenismo in Cuzco, 1905–1945,' *Journal of Tourism and Cultural Change*, 6 (1): 39–56.

Atkinson, B. (2009) E-mail to Evan R. Ward, November 5, 2009.

Bingham, H. (1911a) *Across South America: An Account of a Journey from Buenos Aires to Lima by Way of Potosi*, New York: Houghton Mifflin Company.

—— (1911b) Letter to Louis Schaefer, December 27, 1911, Yale University Library, Manuscripts and Archives, Yale Peruvian Expedition Records (Yale University Library), Box 6 (1911 Nov.–Dec.).

—— (1911c) Letter to Louis Schaefer, April 29, 1911, Yale University Library, Manuscripts and Archives, Yale Peruvian Expedition Records, Box 5 (1909–1910, 1911 Jan.–Oct.).

—— (2003) *Lost City of the Incas*, London: Phoenix Press.

Braniff International (1968) 'The Next Place,' *Life*, January 19, 1968, pp. B2–B3. Online, http://books.google.com/books?id=XUoEAAAAMBAJ&lpg=PA14-IA3&dq=braniff%20international%20machu%20picchu%20ad&pg=PA1#v=onepage&q=braniff%20international%20machu%20picchu%20ad&f=false (accessed January 18, 2012).

Clement, H.G. (1961) *The Future of Tourism in the Pacific and Far East*, Washington, D.C.: U.S. Department of Commerce.

—— (1965) *The Outlook for Tourism in Peru*, Washington, D.C.: Cheechi and Company, pp. 29–30.

COPESCO (c.1973) *La actividad turístico y el desarrollo economico del Perú: un estudio de diagnóstico*, Industrialgrafica S.A., pp. 35, 3; 62.

—— (c.2007) 'Proyecto especial regional: plan Copesco, plan estratégico institutional, 2007–2011.' Online, www.peru.gob.pe/docs/PLANES/13042/PLAN_13042_Plan%20Estrategico%202007%20-%202011_2008.pdf (accessed January 10, 2012).

Defert, P. (1967) *The Development of Peru's Historic Sites and Monuments: A Synthesis of Economic Prospects*, Paris: UNESCO, pp. 3–7.

Defert, P. and Pichot, R. (1988) 'Le tourisme, solution aux dettes de Tiers-monde,' *Revue de tourisme*, 2: 2–5.

Deforges, L. (2000) 'Site Tourism Institutions and Neo-liberal Development: A Case Study of Peru,' *Tourism Geographies*, 2 (2): 178.

D'Harnoncourt, R. (1953) Letter to Louis J. Garcia, November 6, 1953, Museum of Modern Art Archives, New York, René d'Harnoncourt Papers, VIII.24.

El Congreso del Peru (2009) 'Estadísticas de visitantes de los museos y sitios arqueológicos del Peru (1992–2009).' Online, www.congreso.gob.pe/museo/estadisticas.pdf (accessed January 10, 2012).

Engel, F. (1967) *Protection and Exploitation of Historical Monuments*, Paris: UNESCO, p. 8.

Falco, N.J. (1994) 'Inventory/Register to the Records of W.R. Grace,' Archive/Maritime Historical Records Collection, Stephen B. Luce Library, SUNY Maritime College, May 1994. Online, http://sunymaritime.edu/stephenblucelibrary/pdfs/W.%20R.%20Grace%20Papers.pdf (accessed January 10, 2012).

Frischlander, A. (1968) *Acondicionamiento del sitio de Machu-Picchu*, Report for UNESCO, November–December 1967. Online, http://unesdoc.unesco.org/images/0000/000087/008702sb.pdf (accessed July 22, 2016).

Garcia, L.J. (1954) Letter to Rene d'Harnoncourt, February 18, 1954, Museum of Modern Art Archives, New York, RDH VIII.24.

Heaney, C. (2010) *Cradle of Gold: The Story of Hiram Bingham, a Real-Life Indiana Jones, and the Search for Machu Picchu*, New York: Palgrave Macmillan.

Kane, Robert S. (1962) *South America A–Z*, New York: Doubleday, p. 340.

Lombardi, G. (1977) 'Cuzco [*sic*] y su territorio – problemas de planificación,' UNESCO: Centro de Restauracion, Cuzco, pp. 8; 11–12.

Lovón Zavala, G. (1982) *Mito y realidad del turismo en el Cusco*, Cusco: Centro de estudios rurales andinos 'Bartolome de las Casas.'

Lovón Zavala, G. and Moscoso Blanco, R. (1981) *Investigación sobre desarrollo regional Cusco, 1950–1980, 5.7 Informe: sector turismo*, Cusco: Centro de estudios rurales andinos 'Bartolome de las Casas,' pp. 3–5.

MacCannell, D. (1999) *The Tourist: A New Theory of the Leisure Class*, Berkeley: University of California Press.

Mould de Pease, M. (2007) Letter to President Alan Garcia Perez, Online, http://icomos.org/cartapresidente.htm (accessed January 10, 2012).

Panagra (1959) 'You're Not a World Traveler until You Can Name These Eight Places!' *National Geographic*, February 1959, p. 315.

Shankland, G. (1969) *Estudios urbanísticos de Cusco y Machu Picchu*, Paris: UNESCO, pp. 18–19.

Shullenberger, G. (2008) 'That Obscure Object of Desire: Machu Picchu as Myth and Commodity,' *Journal of Latin American Studies*, 17 (3): 317–333.

UNESCO (1966) '1967: International Tourist Year.' *UNESCO Courier*, December 1966, pp. 5–10.

Urry, J. (2009) *The Tourist Gaze* (2nd ed.), London: SAGE Publications.

Villesa Lescano, C. (2007) *Introducción al turismo: teoría y realidad peruana* (8th ed.), Lima: Carlos Villena Lescano.

W.R. Grace & Company (1911) 'Peru – the Land of the Incas,' *National Geographic*, August 1911, p. 787.

14 Shandong Province and tourism

An examination of World Heritage Sites

Ina Freeman and He Sun

Shandong Province of the People's Republic of China hosts two World Heritage Sites: the temple and cemetery of Confucius and the Kong family mansion in the city of Qufu; and Mount Tai (Taishan where 'shan' means mountain), which spans the cities of Tai'an and Jinan. Of the ten criteria for a site to be given the status of a United Nations Educational, Scientific and Cultural Organization (UNESCO) World Heritage Site (WHS), Mount Taishan meets seven and the temple and cemetery of Confucius and the Kong family mansion in Qufu meets three. In terms of these criteria, both represent a masterpiece of human creative genius; include an outstanding example of architecture and landscape illustrating a significant stage in human history; and are directly associated with living traditions and beliefs of outstanding universal significance (UNESCO 2011b). Mount Taishan also has four additional criteria, including: it exhibits an important interchange of human values on landscape design; it bears exceptional testimony to a living cultural tradition; it is an outstanding example of traditional human settlement that is representative of human interaction with the environment; and it contains superlative natural phenomena of aesthetic importance (UNESCO 2011b). To support these sites, the government of the province of Shandong hosts a tourism website, providing information on all tourism sites in the province (Shandong Tourism 2011). This site is written in Mandarin and offers no specialized advertising for the two WH sites. We discuss in this chapter the strategies and tasks that are necessary to increase the benefits of these sites to Shandong.

Background

The temple and cemetery of Confucius and the Kong family mansion in Qufu were first constructed in 478 BC and, following their destruction, were expanded to more than 100 buildings holding more than 100,000 of his descendants (UNESCO 2011a). The Kong family mansion has been added to and now encompasses 12,740 square meters (China Highlights 2011). Confucius was known as a philosopher and teacher with core beliefs on morality and humanness. These beliefs rose and fell in popularity. In the last century, the beliefs were resurrected by Chiang Kai Shek in the 1930s, suppressed during the Mao era, and again promoted in the 1990s (Bell 2008) as a countermeasure to the corruption predominant in China (Zhu 2011). With the Chinese government allocating $10 billion to establish a global network of schools based on the promotion of the Chinese language and culture as influenced by Confucius (Zhu 2011), this philosophy is again back in fashion. Today, there are six million people who identify with Confucianism (ChartsBin 2011b).

Mount Tai, or Taishan, carries significance for both Buddhism and Taoism and has been the destination of pilgrimages for over 3,000 years (UNESCO 2011c). With six million Taoists (ChartsBin 2011c) and 397 million Buddhists (ChartsBin 2011a) in the world, this site has significant appeal. Additionally, the site also notes places that Confucius visited, joining the two locations and adding to the identification of Mount Tai as a symbol of China's national spirit. Here both the natural and the man-made are joined together to symbolize Chinese culture and natural glory.

Both sites are tourism destinations within Shandong Province and both are listed as cultural sites, which have a stronger appeal than natural sites (Yang, Lin and Han 2010). Tourism is recognized globally as a source of funding for development. Tourism is also a means through which the government can promote the unique identity of China and Shandong (Henderson 2002). Mainland China is reported to have hosted 55.98 million overseas tourists in 2010 (APF 2011) with an estimated $45 billion USD (ChinaDaily 2011), although the Beijing Olympics and Paralympics may have artificially inflated those numbers. It is estimated that the number of domestic travelers was 2.12 billion person-times in 2010, resulting in $234.8 billion USD (ChinaDaily 2011). Shandong Province is reported to have 289 million domestic tourists, with revenue of $35.5 USD billion, and 3.1 million overseas tourists, with revenue of $1,765.3 million (Hong Kong Trade Development Council 2010). These tourists are recorded as staying during January to October 2010 a total of 6,682,137 nights, a 17.75 percent growth from 2009 (National Tourism Administration of the People's Republic of China 2011). For tourism to continue to provide the benefits of increasing numbers of tourists, as is predicted by the China Tourism Academy (Yang 2011), some of these receipts need to be funneled back into the maintenance and preservation of the sites as well as ways and means to expand tourism. To this end, the industry is reported to be amalgamating and developing by using modern business techniques, science, and technology to transform and upgrade the tourism sites and products including a national high speed railway (Yang 2011).

These developments may facilitate the improvement of domestic tourism in Shandong Province, which is located in the east of China, along the coast. Shandong has ready access to rail, ship, bus, and air travel, facilitating the movement of inbound tourists, both domestic and foreign, to the World Heritage Sites. The perception of past experiences and cultural tour involvement are particularly important for domestic tourists (Shen et al. 2009).

Tourism and marketing World Heritage Sites

The list of World Heritage Sites includes 911 properties, with 704 cultural, 180 natural, and 27 mixed properties; China boasts 40 properties, with Shandong

Province hosting two (UNESCO 2011d). Mount Tai is one of the mixed sites, of which there are few; in China only four mixed sites are to be found. The World Heritage Sites are significant to tourism in China in enhancing the tourist visit for a large number of tourists (Yang et al. 2010). This study found the increased appeal of the cultural aspects of World Heritage (Yang et al. 2010) significantly increased the numbers of tourists and their length of stay (Yan and Morrison 2007), to the economic benefit of Shandong.

Coupled with the opening of China to foreign visitors, which increases the number of people visiting Mount Tai as well as the Confucian home and temple, is the need to maintain and improve the facilities (du Cros 2007). Unfortunately, the maintenance and improvement of the facilities and the sites increase the number of tourists, potentially resulting in overcrowding (du Cros 2007), an issue first brought to light for the World Heritage Sites in relation to tour groups during public holidays (Cocks 2005 cited in du Cros 2007). The World Tourism Organization has produced a guidebook (UNWTO 2004) to address these issues and reduce the negative impact on the site and the visitor. While this brings to the fore the varying levels of comfort for the tourist in crowded spaces (Kaya and Weber 2002), crowding generally is perceived to have a negative influence on the domestic and foreign tourist (Lawson and Thyne 2001). This is an element that must be addressed by tourism boards, specifically when looking to increase tourism to these sites.

Thus although China is beginning to promote tourist destinations, their hierarchical management, governing bodies, and the Transparency International evaluation of corruption in China in 2015 at 37, where 100 is totally corruption free and a ranking of 83 out of 168 countries (Transparency International 2016), provide hurdles to the promotion and maintenance of tourism sites. The government of China is heavily involved in tourism, which may put people who are inept at marketing or tourism in charge of managing tourism, leaving this potential money making activity underdeveloped for foreign tourists.

One potential solution is the development of e-tourism, using e-marketing as promotion. Information and communication technology (ICT) has had a tremendous impact on the tourism industry in promoting sites that have little available money to compete with well-funded destinations. China has recognized this by instituting a National Tourist Office website (China National Tourism Office 2011b) as well as tourism offices for the provinces. Using these websites, China and its provinces are able to promote their attractions as well as advertise, inform, sell, and facilitate tourism to Mandarin speakers.

Another important aspect of e-marketing and e-tourism is the importance of gathering information from potential and actual tourists via the internet. However, among the 255 million websites as of December 2010 (Pingdom 2011), websites today must contain the correct keywords, be easy to use, be current, and be attractive. A search using Google indicates about 265,000 websites containing the words 'travel, Shandong, tourism' and 141,000 websites using Yahoo (search done 11 March, 2011). If 'World Heritage' is added to the search, Google reports 171,000 sites and Yahoo reports 7,510 sites (search done 11 March 2011). While 16,900 of the latter Google sites and 191 of the latter Yahoo sites are for travel agents, there remains a large number of sites, certainly enough to frustrate the searching public. While China demands businesses structured in China are primarily owned by citizens of China, one wonders if the money that goes to travel agents remains in China where it can facilitate the development of tourism, particularly the World Heritage Sites. This means that for the government of Shandong and China to have tourism websites in Shandong that are able to gather useful information, special efforts must be made to attract the web browser. The information gathered by these websites will potentially inform the tourism bodies and the government of the needs and wants of the interested tourist.

Information on foreign tourists is especially important for the Shandong tourism authorities because these tourists bring money into the country. This influx of money will help the economy of China by increasing the real revenue of the country. This revenue can then be spent on the maintenance and development of tourism and tourism sites, without diverting internal revenues. However, to capture the information, web based communication and capture tools must be improved and coordinated, potentially by the National Tourism Authority, to build an authentic and trustworthy source of information.

Much of the information on the internet about the World Heritage Sites and tourism in Shandong is in Mandarin, as found on the primary website for the Shandong Tourism Authority (Shandong Tourism 2011a), thus excluding non-Mandarin speakers. This site provides links to two additional pages, one for general tourism (Shandong Tourism 2011b) and one for more on the cultural aspects of tourism (Shandong Tourism 2011c). The first page for general tourism further provides links via flags to webpages in English (Shandong Tourism 2011d) and the Japanese page (Shandong Tourism 2011e). The use of websites allows the Shandong Tourism Authority to establish relationships with consumers, an important aspect for an industry that must cater to the individual needs and wants of tourists and the tourism industry inside and outside China.

Domestic tourism is important, as can be seen from the number of websites for Shandong Province that are specifically in Chinese as well as the national government instituted three weeks a year that are individually classified as a 'gold week.' These weeks encompass May Day (May 1), the National Anniversary Day (October 1), and Spring Festival or Lunar New Year (usually at the end of January or in the first part of February). The importance of these days to tourism is reflected in 2008 estimates that 178 million tourists traveled then, representing a 22 percent increase on the previous year (National Tourism Administration of the People's Republic of China 2008). However, this same report does not list the cities around the World Heritage Sites among the major tourism cities included in the national holiday tourism statistics and forecasting system. This represents a significant loss for Shandong Province.

Current tourism in Shandong Province

Shandong Province's tourism is guided by the governmental tourism agency Shandong Travel Administration. All four websites are structured differently. All are authorized by the government and target different interests. The first Chinese site (Shandong Tourism 2011b) was updated following the sharing of the Freeman and Sun (2010) findings, which made numerous recommendations concerning the inadequacies of the cultural sites, with officials of Shandong Travel Authority.

This site links to 'friendly Shandong,' 'charming Shandong,' 'travelling in Shandong,' 'wonderful activities,' 'visual experience,' 'discount zone,' and 'online tourism bureau.' On the left of the homepage is a link called 'Tourist's Tools' which gives information and reservations for hotels, restaurants, and travel. Other information includes links to tourism agencies and activities such as entertainment, shopping, tourism sites, and physical activities. Other links provide duplicate information found under 'wonderful activities' and 'discount zone' as well as comments from other tourists. Information concerning local and national tourism-related policies, complaints procedures, and other useful information is also included. Weather forecasts and the provision of an e-map are also linked in, although the e-map is currently under construction. Furthermore, online help is available via e-mail. Under 'friendly Shandong' the World Heritage sites are discussed with detailed descriptions. Further useful information is provided on the other 37 tourism sites, shopping, entertainment, lodging, and transportation. 'Charming Shandong' links to a map of Shandong, highlighting the 17 major cities. 'Traveling Shandong' links to different types of tours, including cultural, natural, business, themed, leisure, and general tours. 'Wonderful activities' links to information about Chinese New Year celebrations. 'Visual experience' contains photos and videos of the province. 'Discount zone' gives information about discounts. 'Online tourism bureau' links to the official website of Shandong Tourism Bureau.

The second site (Shandong Tourism 2011c) exhibits a creative design but has a number of problems such as providing information that is more than a year out of date, blank tabs and malfunctioning links, inadequate information in many of the links, and no significant emphasis on the World Heritage Sites among the 37 other tourism sites in Shandong Province. Much of the information on the second site is similar to that provided on the first site. While offering some multimedia information specifically on the culture and heritage of the tourism sites, the inoperability of the site and the dated information detracts from its usability.

The Japanese site opens to many pictures, including Mount Tai and Confucius temple and cemetery, a travel map with information concerning flights, tourism agencies, sightseeing, restaurants, shopping, and activities. The landing page also provides a link to a welcoming message. Although there is not an extensive amount of information on this site, it appears to be current.

The site in English is fraught with problems. The front page highlights a 'Ride to Confucius' sponsored by Harley-Davidson in 2008. While this site carries the tourism logo of Shandong Province (Friendly Shandong), the most current information is the landing page, dated October 2008. Some linked pages are blank (e.g. the tourism routes and themed tourism), while others contain information from 2005 (destination cities). The links to restaurants, hotels, travel agencies, scenic spots, shopping, and entertainment are not functional, leaving the site visitor with little access to information. The links to the e-journal date from 2007 and the search engine takes you to a Chinese Google search page. Clearly neither website provides all the information that is optimal for the facilitation of inbound tourism.

Although the China National Tourism Office (2011b) website is not dated, it contains a mix of current and dated information. While it does not mention Confucius' home, it does mention Mount Tai, although not as a World Heritage Site. Similarly with the government sponsored China Travel Guide published by the National Tourism Administration, Shandong Province is discussed, including the two World Heritage Sites, although neither is mentioned as such (China Tour.com n.d.). However, this is not to say that the Chinese government ignores World Heritage Sites, as indicated in the government's publication of their nomination of 54 additional sites (World Heritage Site 2011).

This lack of attention is not shared by China.org.cn, a website that partners with other news sources targeting multiple languages. Shandong has a separate page called 'Charming Shandong, Birthplace of Confucianism' (China.org.cn 2011). Multiple aspects of Shandong are discussed, including the economy, society, infrastructure, history and culture, natural resources, as well as tourism, where the World Heritage Sites are specifically mentioned as well as other tourism sites. China.org.cn also provides a page that lists the World Heritage Sites throughout China and gives information on all of them (China.org.cn n.d.). This site offers links to travel agencies, suggestions for destinations and tours, updated events and festivals, and useful information for the traveler, including currency converters, a quick language guide, visa and customs information, train timetables, an airline guide, and travel agencies. This news agency also provides links to government statistics concerning international tourists provided by the National Tourism Administration of the People's Republic of China, as well as national laws that are pertinent to the international traveler to China (China.org.cn 2011).

Thus, despite Shandong being noted as one of the most populous and affluent provinces in China (China Tour.com n.d.), its global communication about tourism in general and the World Heritage Sites in particular is lacking in currency, detail, and depth in any languages other than Mandarin and Japanese. There are numerous tourism websites in multiple other languages which do bring tourists to China and provide some income. However, given the reduced ability to draw external tourists and restricted income, the tourism industry in Shandong Province may ultimately be harmed.

Possible solutions

With the resurgence in the government's acceptance of Confucian thought, as indicated by the institution of a new peace prize in December 2010 bearing the name the Confucius Prize (MacLeod 2010) and the growth of Confucian thought with the institution of Chinese and Confucius classroom programs around the world (Welch 2011), the World Heritage Sites in Shandong Province have increasing importance in the world. However, when searching for information about tourism specifically designed for these, the Shandong Tourism Administration English website is lacking. While the site in Mandarin offers information, there is little information on one Chinese site that links the World Heritage Sites with the facilities and amenities posted. The lack of easily accessible information for the occidental world is reflected in the nationalities of the tourists who visit Shandong Province; most originate from Korea (41 percent) and Japan (18 percent), followed by Hong Kong, Macao, Taiwan, Singapore, and Malaysia (Shandong Province Statistical Yearbook 2009).

Despite the fact that the economy of Shandong has grown, boasting a lead in some of the major economic indicators (China.org.cn 2011), its promotion of its two World Heritage Sites on the English and second Chinese website does not reflect this. Because e-marketing reaches a global audience and the foreign tourist is the individual who brings money into the country, the official tourism agency, Shandong Tourism Administration, must create and maintain current, accurate, and inclusive websites. That is, the links must actually take the browser to an active page and the website must be updated regularly. The two Chinese websites should be linked, and the information should be distinctly different.

The websites could be a one-stop shop that provides one-click accessibility to information on restaurants, hotels, museums, tours, and other activities such as booking tickets and reservations for rooms or vehicles. Here photography could be included to make it possible to view the hotel and the room, or the restaurant meals, or the actual buses used. Another aspect the website might include is a differentiation of hotels by stars or categories, such as bed-and-breakfast, and restaurants by the type of food provided (authentic Chinese, Western, and others). This facilitates the planning necessary for hotels, restaurants, tours, and other activities, providing a smooth provision of services to the tourist, thereby encouraging repeat visits.

The site provides links to tourism operators, promoting the cooperation of regional tourism by creating a demand and supply chain of tourism. By facilitating and reinforcing these linkages, the site could also facilitate the coordination of tourism activities into comprehensive tours. For example, tour buses could arrange with local restaurants to stop for meals or coordinate with hotels to provide lower rates for tour participants in return for the tour starting at the hotel. To facilitate the traveler, a link to accurate map pages could be made, allowing the visitor to plan activities to optimize their time in Shandong Province.

The 'Friendly Shandong' brand could also be expanded through all those internal and external to Shandong Province involved in tourism. The brand appears on all the websites and is in both Mandarin and English. The promotion of this brand will enhance the competitiveness of Shandong tourism within the global tourism environment. The linkage of this brand with sustainable tourism could further enhance the brand globally and assist in the sustainable development of tourism in the province.

The building of tourism will require an examination of the home country of the foreign tourists. While many tourists originate from the Asiatic region close to

China, there are few visitors from Europe, North and South America, the Middle East, and Africa. Furthermore, the discrepancy in the previously discussed numbers of domestic to foreign tourists remains a problem for an industry that wants to attract additional income into the country. These foreign tourists stay for a longer period of time, due to the time spent traveling. These markets are the ones where the websites are most important for planning and information, yet the website that is in English is outdated.

To facilitate more visits from these tourists, the websites must all be updated. This requires funding for a webmaster; the funding can come from charging those who list their businesses on the site. The small and medium enterprises would also be promoted globally to tourists, making them choice destinations. The linking of other organizations' sites, such as the National Tourism Administration of the People's Republic of China, the UNESCO World Heritage Sites, and China.org.cn, to provide current news and weather would add to the credibility of Shandong Tourism Administration as a 'one-stop-shop.'

E-marketing not only enables the establishment of relationships with customers from the four corners of the earth but also facilitates the maintenance and enrichment of these relationships. Through these relationships, foreign tourist spending will increase the real revenue of the country. This money could be used to enhance and maintain the website, providing a means through which Shandong Tourism Administration can promote the province at little or no cost to the government.

The employment of a webmaster to orchestrate tourism in Shandong Province could highlight specifically the World Heritage Sites while providing information on other sites and on Shandong Province in general. The webmaster will also be able to finesse the use of keywords to facilitate the site being found using common search engines, including Baidu, the leader in China. This includes ensuring the provision of translated subpages in multiple languages. With the added traffic, the webmaster will need to continually monitor the site to ensure its integrity.

Other responsibilities within this position include the gathering of data from web browsers and site utilizers, including the number of visitors, their activity, and their country of origin. Recording the popularity of each page will facilitate the construction of a website that responds to users, making the site more popular. Users can be encouraged to bookmark the site to access continually updated information. Feedback links will request demographic and other information such as intentions for visiting Shandong as well as e-mail addresses, providing information for a database that can be used for promotional campaigns.

The provision of virtual tourism through these sites also creates links to individuals contemplating a visit. Using the current technology of Web 2.0 and the emerging Web 3.0, Shandong Tourism can embed videos, blogging sites, transactional information sites, and other tools to encourage visitors. Communication via Web 2.0 and the emerging Web 3.0 can be interactive, but it must be regularly monitored and coordinated, a role Shandong Tourism Administration should assume as the primary authority on tourism in the province.

This communication can facilitate the delivery of feedback or evaluations of the website, tourism destinations, policies, and the collection of information, all of which is missing from the Shandong Tourism Administration website. While many of the problems of the latter website can be circumvented by using travel agents, a lot of international tourists prefer to use the internet for travel and tourism planning (Ataeefar and Teymoori 2008). If the websites are non-functional, tourism will be limited.

Conclusion

The growing interest in China due to the opening of its doors to foreign visitors and China's emergence as a dynamic economy are precursors to increased tourism to China. The growing interest in Confucius and the recognition of Mount Tai as a dual World Heritage Site are factors that will further increase tourism (Yang et al. 2010) to Shandong.

On examining the tourism of Shandong Province with reference to the World Heritage Sites of Confucius and the Kong family mansion and Mount Tai, it is apparent Shandong province is not optimizing the tools available to encourage tourism, specifically for these sites. Tourism is an industry that spreads money significantly throughout the population. With the increase in tourism, Shandong will continue to be an economic leader in China. An increase in tourism necessitates an increase in the number of hotels, restaurants, tour companies, and retailers, meaning an increase in government revenues from taxation and higher employment from construction and the provision of services.

By using the internet, Shandong Tourism Administration can begin to build its foreign tourism appeal and market itself as a destination of choice. This will not only benefit the individual sites, but also facilitate the development of Shandong Province. It is time to close the divergence between theory and practice in Shandong's tourism (Sun 2008) by implementing changes. Thus, the task of the local government of Shandong and Shandong Tourism Administration is to improve the pre-existing tools in order to attract visitors to the World Heritage Sites and so garner the money to pay for the maintenance and conservation of the sites.

References

APF (2011) China World's Third Most Visited Country: UN. Online, www.google.com/hostednews/afp/article/ALeqM5hFli5xB9YgVoSunOYYFREW6BozmQ?docId=CNG.f55f656a9f597ee071fe1ead97d63e4a.3b1 (accessed 8 March 2011).

Ataeefar, Z. and Teymoori, T. (2008) 'The Impact of E-tourism in Developing Countries: A Case for Cambodia', Third International Conference on E-commerce with Focus on Developing Countries, 23 October 2008, Isfahan, Iran.

Bell, D.A. (2008) *China's New Confucianism: Politics and Everyday Life in a Changing Society*, Princeton, NJ: Princeton University Press.

ChartsBin (2011a) Buddhism Adherents by Country. Online, http://chartsbin.com/view/wq1 (accessed 10 March 2011).

ChartsBin (2011b) Confucianism Adherents by Country. Online, http://chartsbin.com/view/6uq (accessed 10 March 2011).

ChartsBin (2011c) Taoism Adherents by Country. Online, http://chartsbin.com/view/pog (accessed 10 March 2011).

ChinaDaily (2011) 'China's Tourism Revenue up 20% in 2010'. Online, www.chinadaily.com.cn/business/2011-01/13/content_11844956.htm (accessed 8 March 2011).

China Highlights (2011) Kong Family Mansion. Online, www.chinahighlights.com/qufu/attraction/kong-family-mansion.htm> (accessed 8 March 2011).

China National Tourism Office (2011a). Landing Page. Online, www.china.org.cn/china/shandong/node_7087752.htm (accessed 12 March 2011).

China National Tourism Office (2011b). Landing Page. Online, www.tourismchina.org/ (accessed 12 March 2011).

China.org.cn (2011) Charming Shandong, Birthplace of Confucianism. Online, www.china.org.cn/china/shandong/node_7087752.htm (accessed 12 March 2011).

China.org.cn (n.d.) World Heritage Preservation in China. Online, www.china.org.cn/english/whp/71256.htm (accessed 11 March 2011).

China Tour.com (n.d.) Shandong. Online, www.chinatour.com/newpage/cities.htm# (accessed 10 March 2011).

du Cros, H. (2007) 'Too Much of a Good Thing? Visitor Congestion Management Issues for Popular World Heritage Tourist Attractions', *Journal of Heritage Tourism*, 2(3): 225–238.

Freeman, I. and Sun, H. (2010) 'Promoting the Cultural and Natural Heritage Sites in the Shandong Province, China', World Heritage and Tourism: Managing for the Global and the Local, 2–4 June 2010, Quebec City, Canada.

Henderson, J. (2002) 'Heritage Attractions and Tourism Development in Asia: A Comparative Study of Hong Kong and Singapore', *International Journal of Tourism Research*, 4(5): 337–344.

Hong Kong Trade Development Council (2010) Shandong Province Major Economic Indicators. Online, www.hktdc.com/info/mi/a/mpcn/en/1X06BVNS/1/Profiles-Of-China-Provinces-Cities-And-Industrial-Parks/SHANDONG-PROVINCE.htm (accessed 8 March 2011).

Kaya, N. and Weber, M.J. (2002) 'Cross-Cultural Differences in the Perception of Crowding and Privacy Regulation: American and Turkish Students', *Journal of Environmental Psychology*, 23(3): 301–309.

Lawson, R. and Thyne, M. (2001) 'Destination Avoidance and Inept Destination Sets', *Journal of Vacation Marketing*, 7(3): 199–208.

MacLeod, C. (2010) 'China Counters Nobel with Confucius Prize', *USA Today*, 12 December 2010. Online, www.usatoday.com/news/world/2010-12-10-chinanobel10_ST_N.htm (accessed 11 March 2011).

National Tourism Administration of the People's Republic of China (2008) Tourism Statistics Report of the 'Oct. 1' Gold Week 2008. Online, http://en.cnta.gov.cn/html/2008-11/2008-11-9-21-37-90694.html (accessed 11 March 2011).

National Tourism Administration of the People's Republic of China (2011) International Tourists by Locality, Jan.–Oct. 2010 (no. 2). Online, www.cnta.gov.cn/html/2011-1/2011-1-5-15-40-45294.html (accessed 11 March 2011).

Pingdom (2011). Internet 2010 in Numbers. Online, http://royal.pingdom.com/2011/01/12/internet-2010-in-numbers/ (accessed 10 March 2011).

Shandong Province Statistical Yearbook (2009) *Yearbook 1991–2008*, Beijing: China Statistic Publishing House.

Shandong Tourism (2011) Online, www.shandong.travel/ (accessed 8 March 2011).

Shandong Tourism (2011a) Landing Page. Online, www.sdta.gov.cn/ (accessed 8 March 2011).

Shandong Tourism (2011b) Landing Page. Online, www.sdta.cn/dtss/ (accessed 8 March 2011).

Shandong Tourism (2011c) Landing Page. Online, www.shandong.travel (accessed 8 March 2011).

Shandong Tourism (2011d) Landing Page. Online, www.travelshandong.us (accessed 8 March 2011).

Shandong Tourism (2011e) Landing Page. Online, www.tabi-shandong.com (accessed 8 March 2011).

Shen, S., Schüttemeyer, A. and Braun, B. (2009) 'Visitors' Intention to Visit World Cultural Heritage Sites: An Empirical Study of Suzhou, China', *Journal of Travel and Tourism Marketing*, 26(7): 722–734.

Sun, L. (2008) 'Societal Transition: New Issues in the Field of the Sociology of Development', *Modern China*, 34(1): 88–113.

Transparency International (2016) Corruption Perceptions Index 2015 Results. Online, www.transparency.org/cpi2015#downloads (accessed 13 August 2016).

UNESCO (2011a) Temple and Cemetery of Confucius and the Kong Family Mansion in Qufu. World Heritage Convention. Online, http://whc.UNESCO.org/en/list/704 (accessed 8 March 2011).

UNESCO (2011b) The Criteria for Selection. World Heritage Convention. Online, http://whc.UNESCO.org/en/criteria> (accessed 8 March 2011).

UNESCO (2011c) Mount Taishan. Online, http://whc.UNESCO.org/en/list/437> (accessed 8 March 2011).

UNESCO (2011d) World Heritage List. Online, http://whc.UNESCO.org/en/list (accessed 10 March 2011).

UNWTO (2004) *Tourism Congestion Management at Natural and Cultural Sites*, Madrid: World Tourism Organization.

Welch, C. (2011) 'China-Sponsored Language Programs in U.S. Raise Concerns, Hopes', *CNN U.S.* Online, www.cnn.com/2011/US/01/19/china.funds.language.programs/index.html (accessed 11 March 2011).

World Heritage Site (2011) Landing Page. Online, www.worldheritagesite.org/countries/china.html (accessed 10 March 2011).

Yan, C. and Morrison, A.M. (2007) 'The Influence of Visitors' Awareness of World Heritage Listings: A Case Study of Huangshan, Xidi and Hongcun in Southern Anhui, China', *Journal of Heritage Tourism*, 2(3): 184–195.

Yang, C.-H., Lin, H.-L. and Han, C.-C. (2010) 'Analysis of International Tourist Arrivals in China: The Role of World Heritage Sites', *Tourism Management*, 31(6): 827–837.

Yang, G. (2011) China's Tourism Sector Realizes Fast Growth: CTA. *China Economic Net*. Online, http://en.ce.cn/Insight/201101/27/t20110127_22177116.shtml (accessed 10 March 2011).

Zhu, Y. (2011) 'The Confucian Tradition and Chinese Television Today', *New York Times*. Online, www.nytimes.com/ref/college/coll-china-media-003.html (accessed 8 March 2011).

15 The valuation of protected areas

Tourists in Chitwan National Park, Nepal

Jennifer Michelle Cook and Michal J. Bardecki

Managers of protected areas are generally confronted with a classic financial problem: they may be in charge of highly valued properties and yet not have the means to generate an income flow to sustain the assets. In the developing world, sustainable funding resources for conservation are scarce. A lack of adequate funds can jeopardize conservation efforts in protected areas by limiting the ability of managers to carry out critical tasks such as boundary enforcement, species protection and local education (Dixon and Sherman 1990; Navrud and Mungatana 1994; Baral et al. 2008). A systematic valuation can provide protected area managers with a structured approach to appropriate prices for management purposes and indicating ways of capturing the value associated with the areas under their supervision (Pearce and Moran 1994), as well as helping justify appeals for funding support.

A conventional economic approach to evaluate the 'best' use of resources in protected areas is assessment of the economic value of the resource (ICEM 2003). Problematically, many of the economic, social and environmental benefits associated with protected areas have no formal market, price or expressed cash value (Dixon and Sherman 1990). Certain goods and services associated with protected areas can be commercialized; however, if, for example, entry fees and prices for use are administratively set, they may not reflect the true value of the area.

The concept of total economic value (TEV) is now commonly used to evaluate the full scope of benefits and costs of protected areas (Fromm 2000; ICEM 2003). Emerging in the mid-1980s, TEV is now widely used to identify the economic benefits associated with protected areas (CNPPA 1995). Instead of solely focusing on direct commercial values, TEV encompasses the non-market values including ecological functions and non-use benefits associated with protected areas such as bequest values and existence values (Aylward and Barbier 1992; Rietbergen-McCracken and Abaza 2000; Navrud and Mungatana 1994).

Many methods for valuing protected areas have been developed. Various manuals, studies and overviews on the application of TEV to protected areas and ecosystems are available (Dixon and Sherman 1990; Pearce 1992; Pearce and

Moran 1994; CNPPA 1995; Barbier et al. 1997; Rietbergen-McCracken and Abaza 2000; Samdin 2008). These outline a wide range of methods for valuing the range of values associated with protected areas.

The measurement of those values which are not market-based can be challenging. One tool which has gained wide acceptance for researching specific forms of valuation is the contingent valuation method (CVM).

The contingent valuation method

CVM uses a survey to measure individuals' maximum willingness to pay for different natural resources or other public goods presented to them in a hypothetical market (Mitchell and Carson 1989). The method creates a hypothetical market place in which respondents are given the opportunity to 'buy' or 'bid' for the good or service (Whittington et al. 1990). Asking people, as opposed to solely determining values of goods based on market data or 'use value', can assess non-use values or existence values (Hanemann 1994). CVM is widely used as one of the standard approaches to measure the economic values of non-market goods and services such as those associated with recreation and tourism and environmental quality. The method has received official recognition as a recommended valuation technique (Bateman et al. 1999; Ahmed and Gotoh 2006).

Despite criticisms (e.g. Walsh et al. 1984; Hausman 1993; Tietenberg 1996; Mitchell and Carson 1989; Hanemann 1994; Shultz et al. 1998), CVM has been widely used in academic and non-academic research throughout developed and developing countries (Carson et al. 1995; Mitchell and Carson 1989). Notably, CVM has been used to determine visitors' and tourists' willingness to pay for entrance into national parks and protected areas throughout the world, including those in the developing world (e.g. Baral et al. 2008; Day 2000; ICEM 2003; Israngkura 1997; Lee and Han 2002; Mmopelwa et al. 2007; Shultz et al. 1998; Walpole et al. 2001).

Nepal and Chitwan National Park

Nepal is a small, mountainous, landlocked country in the central Himalaya. It is bordered by China to the north and India to the south. Nepal is among the least developed of nations. The greater part of the population is rural; agriculture employs 76 per cent of the workforce, contributing 40 per cent of Nepal's GDP (CBS 2008).

Tourism has been incorporated into the national development strategy in Nepal since the 1970s (DNPWC 2008). Tourism, specifically ecotourism, is seen as a potential source of employment and foreign currency and a way to achieve sustainable economic development. Tourism development has been considered a viable model to develop remote areas: commoditizing the natural beauty of the country and traditional Nepali ways of life and artistic heritage.

Nepal's distinctive geographical position, as well as its altitude and climatic variation, give the country a unique and varied biodiversity (Wells and Sharma

1998). This diversity is found through three distinctive ecological zones: the low-lying tropical Terai, the deciduous and coniferous forests of the subtropical and temperate Hills, and the sub-alpine and alpine Himalaya (WWF Nepal 2006). Over 19.7 per cent (28,998 km^2) of the total area of the country is now declared protected (DNPWC 2010). These areas include nine national parks, three wildlife reserves, one hunting reserve, four conservation areas and eleven buffer zones representing all three ecological zones. The national parks and protected areas in Nepal are managed by the Department of National Parks and Wildlife Conservation (DNPWC), Ministry of Forests and Soil Conservation. The overall goal of the Department is to conserve and manage the rich and varied biological diversity of Nepal with an emphasis on wildlife and protected areas (DNPWC 2010). Many of the protected areas in Nepal are major tourist destinations. The national Department of National Parks and Wildlife Conservation reports that in 2007 the collective revenues from all protected areas were 111,871,108.00 NPR, equalling over 1.5 million USD (DNPWC 2008).

In terms of biodiversity, the Terai is the richest part of the country. Its lowland tropical forests span the Indian–Nepali border. The area is also heavily populated, resulting in high pressure on forest and agricultural resources. The Terai Arc Landscape (TAL) Project is one effort to address the issues of the region. It is a notable landscape-level programme involving a very large geographical area, implementing a wide variety of measures to promote conservation along with the well-being of local people (TAL-WWF 2010). The TAL includes eleven Nepalese and Indian-border protected areas. Important among these, and exemplifying many of the region's issues related to wildlife conservation, environmental problems and population pressure, is Nepal's first national park, Chitwan National Park.

Chitwan National Park, established in 1973, lies along the southern border of Nepal with India at an elevation of 300 m and below. The United Nations Educational, Scientific and Cultural Organization (UNESCO) designated the park as a World Heritage Site in November 1984, recognizing its unique biological resources. The national park covers 1182 km^2 and the surrounding buffer zones encompass an additional 766 km^2.

The biological richness of the park is exceptional, including the endangered Asian one-horned rhinoceros (*Rhinoceros unicornis*), Asiatic elephant (*Elephas maximus*), Bengal tiger (*Panthera tigris tigris*), gaur (*Bos guarus*), Gangetic dolphin (*Platanista gangetica*), giant hornbill (*Buceros bicornis*), Bengal florican (*Houbaropsis bengalensis*) and gharial (*Gavialis gangeticus*) (UNESCO 2003; WWF Nepal 2006; DNPWC 2008; Straede and Treue 2006).

The management challenges in Chitwan National Park are many. The pressure to collect products in the park continues to be strong as 'there is still a gap between local people's needs for supplementing natural resources and their rights to satisfy them on a legal basis' (Straede and Treue 2006: 251). Park–people conflicts, some related to the relocation of people from the park, continue (Luzinda 2008; Straede and Treue 2006; Nepal 2002; Brockington and Igoe 2006; Weladji and Tchamba 2003) and include crop losses to wildlife, attacks by elephant, tiger, leopard or other animals, and flooding from the Rapti River (McLean and Straede 2003). Even though conflicts exist, the results from recent research suggest that a large majority of local residents are generally in favour of the park and believe that it is beneficial to have the national park (Spiteri and Nepal 2008a; McLean and Straede 2003).

The generally unregulated tourism development in and around Chitwan National Park has resulted in various adverse environmental impacts (UNESCO 2003; WWF Nepal 2006; DNPWC 2008). Management plans are weak and generally not properly implemented in most of the Terai's protected areas, Chitwan National Park included (WWF Nepal 2006).

Of all the national parks and protected areas in the country, Chitwan National Park receives the highest number of visitors (DNPWC 2010). More than 55 per cent of these are foreign tourists (i.e. not domestic tourists, nor tourists from South Asia Association for Regional Cooperation – SAARC – nations; see Table 15.1). As with other Nepali protected areas, each category of tourist pays a different price for a single-day entry to the park: 20 NPR for domestic, 200 NPR for SAARC and 500 NPR for foreign tourists (i.e. approximately 0.28 USD, 2.80 USD and 7.00 USD, respectively).

Tourists are attracted to the park by activities such as elephant safaris in both the park and neighbouring community forests, canoeing for sightseeing and bird watching, guided jungle walks, visits to facilities such as the wildlife breeding projects for Asiatic elephants and gharial, jeep safaris inside the park, and watch-tower overnight adventures. Tourist numbers to Chitwan National Park have been rising. In 1977, the park had only one lodge with approximately 5,000 annual visitors. By 2000 there were seven lodges operating inside the park and more than 70 lodges outside the park, predominantly based in the town of Sauraha (Whelpton 2005; Bhusal 2007). There are 22 tourist guide offices with 383 professional guides based in the Sauraha area (Bhusal 2007). Bhattarai et al. (2005) have been critical of Chitwan National Park, describing it as a typical example of externally induced, uneven development rather than a model of wildlife conservation. Several authors argue that the endangered species in the park are indeed protected, but only to be commoditized for the consumption of high-class tourism (Bookbinder et al. 1998; Jones 2007; Spiteri and Nepal 2008b).

Table 15.1 Tourist entrance fees in the Nepali year 2065 (April 13, 2008 to April 15, 2009), Chitwan National Park

Type of tourist	*Number*	*Entry fee (NPR)*	*Total fee amount (NPR)*	*Total fee amount (USD)*	*Percentage of total fees collected*
Foreigner	46,491	500	23,245,500	325,437	90.27%
SAARC	9,815	200	1,963,000	27,482	7.62%
Nepali	27,158	20	543,160	7,604	2.11%
Total	**83,464**		**25,751,660**	**360,523.24**	**100.00%**

Source: Chitwan National Park Ticket Office, Sauraha.

Significant numbers of tourists come to the park and stay in the area for three or four days as part of guided tour groups with all their activities booked in advance. Tourists who are not part of a group may book pre-packaged three or four day trips; these are readily available in the major urban centres of Kathmandu and Pokhara. A third type of tourist travels independently to the park, booking activities through independent agencies selling their services around the park border (Mueller-Boeker 2000).

Park funding is generated through the entrance fees collected at the park. Funds for management of the park are allocated by the government of Nepal. Fifty per cent of the total revenues is earmarked for the central treasury, while 50 per cent is made available at the park level to be used in buffer zone management programmes for local communities. Currently, these funds are not directly available for park management (UNESCO 2003). Budgetary requirements are estimated annually and submitted to the central government for allocation of funds. The actual release of funds is very low and varies from year to year (UNESCO 2003). Funding continues to be a problem for effective management and conservation at Chitwan National Park.

With the many tourism activities and cultural sites around the park, there is a flow of economic benefits to local communities adjacent to the park through tourism and employment opportunities. Bookbinder and colleagues found that 72 per cent of hotel employees and 74 per cent of nature guides around Chitwan National Park originated from the local area, but that the local area was losing substantial tourism dollars because of the high proportion of non-local hotel ownership, some 61 per cent of the total (Bookbinder et al. 1998). Another major economic loss occurs because more than half of the hotel bookings are made in advance in Kathmandu, Pokhara or through an international tourism agency. As a result, significant revenues do not go into the local economy. Positive benefits are also reduced by deflated market conditions resulting from price competition among the many budget hotels in the area.

Research objectives

The aim of this study is to explore means to further conservation and development efforts at Chitwan National Park by providing insight into whether park managers can better capture value from foreign tourists visiting the area. A contingent valuation tool is used to explore the valuation associated with Chitwan National Park. Specifically, the research objectives are to:

1. determine foreign tourists' willingness to pay while visiting Chitwan National Park and the motivation for their visits;
2. evaluate those visitor characteristics which are significantly related to tourists' willingness to pay; and
3. explore whether or not the potential exists to capture value from tourists through changes in current entrance fees.

Method

Between October 16 and November 12, 2009 a questionnaire was administered to a convenience sample of foreign tourists staying in accommodations in Sauraha, Nepal. At the time seven resorts were located within Chitwan National Park; these tended to cater to those travelling on expensive tours. Initially the plan was to include guests staying within the park; however, all seven of these facilities had had operations (temporarily) suspended in July 2009 following a controversy over the renewal of their contracts – generally, their bookings had been distributed to nearby hotels and lodges. Virtually all tourists visiting the park in the study period stayed in Sauraha.

As a result of the higher fees charged to them, foreign tourists provide over 90 per cent of the total fees collected annually (Table 15.1). This study focuses on this segment, which has the greatest effect on park revenues.

The questionnaire generally parallels those from the two recent case studies in the Annapurna region of Nepal (Baral et al. 2008; Thomson 2007). It consisted of both open-ended and closed questions addressing:

1. the primary purpose respondents had for coming to Nepal and details about their travel arrangements;
2. their motivations in visiting Chitwan National Park;
3. their perceptions of tourism in the Chitwan National Park area;
4. their general attitudes towards conservation and ecotourism;
5. their expenditures and knowledge of the park entry fees;
6. their general socio-economic characteristics (e.g. age, sex, nationality, level of education, employment status, annual expenditure on recreation and tourism);
7. their willingness to pay for entry into Chitwan National Park.

The willingness to pay portion of the questionnaire followed norms of the contingent valuation procedure (Arrow et al. 1993; Tietenberg 1996). Questioning was done orally after respondents had completed the rest of the questionnaire. Respondents were scaled through high and low bids (a bidding game elicitation technique) to find the highest point at which they were willing to pay, starting with an initial 'bid' of 7 USD (the current entrance fee for foreign tourists). Follow-up questions probed the reasons for the final 'bid'. In total 203 questionnaires were successfully completed.

Results and discussion

Profile of respondents

Of the 203 people interviewed, 52.7 per cent were women and 47.3 per cent were men – a slightly higher proportion of women than recorded by the Chitwan National Park Ticket Office in Sauraha (49.5 per cent). The respondents' mean age was 38.2 years old (median = 35 years) with a range from 14 to 71 years. The respondents were generally highly educated: 77.5 per cent reported completing at least an undergraduate level of education.

The sample included visitors from 23 different countries: 75.4 per cent were from Europe (UK: 22.2 per cent, the Netherlands: 17.2 per cent, Germany: 15.8 per cent), 13.3 per cent from North America, 7.4 per cent from Oceania, and 2.5 per cent from Asia (non-Nepali and non-SAARC tourists). These proportions differ significantly from those of all foreign tourist arrivals in Nepal (GoN 2008) by including greater proportions of Europeans and fewer Asians. One issue with this study was the limited means of access to organized tour groups of Chinese tourists and language barriers with this segment – they are under-represented. Respondents were asked to estimate their total annual expenditures on recreation and tourism. The modal category was 2,000–5,000 USD; almost one in five spent in excess of 10,000 USD.

Trips to Chitwan National Park were short (the median was three days) and were generally organized as part of a more extended tour or trip in Nepal alone or including neighbouring India. In addition, a minority had been in Kathmandu for reasons other than tourism and had taken advantage of tour companies' promotion of Chitwan National Park as a 'quick trip down to the jungle'.

The respondents were about evenly divided between those (49.8 per cent) who were independent travellers and the remainder who had made their arrangements through a group or agent. Of those travelling in an organized tour, 75.8 per cent had made their travel arrangements to the park through a travel agent or travel company while in their home country.

Those travelling in an organized group reported that their park fees had been paid through their travel group. Upon being told the current entrance fee was approximately 7 USD, most were surprised at how low it was. Even for those travelling independently, the fee is usually paid as part of an arranged elephant ride or jungle safari and included in the booking price; only 10 per cent of the respondents reported having directly paid their own entry fee at the park. Of the independent travellers, 39 per cent indicated that they did not know the price of the entrance fee.

Contingent valuation

The majority of respondents (N = 186) provided a categorized response concerning their willingness to pay for entry to the park, i.e., 'Yes, I would pay more', 'No, I would not pay more' or 'No, I will not pay the current price'. Over 70 per cent (71.5 per cent) of those who responded indicated a willingness to pay more than the current entry fee; 24.2 per cent would only pay the current fee; 4.3 per cent wanted a lower fee.

The mean willingness to pay for entrance fees at Chitwan National Park among those who fully responded to the contingent valuation survey (N = 124) is considerably higher than the current entrance fees. The respondents' mean willingness to pay is 23.35 USD (median = 14.00 USD; standard deviation = 21.03 USD) with a range from <7 to 100 USD (Figure 15.1).

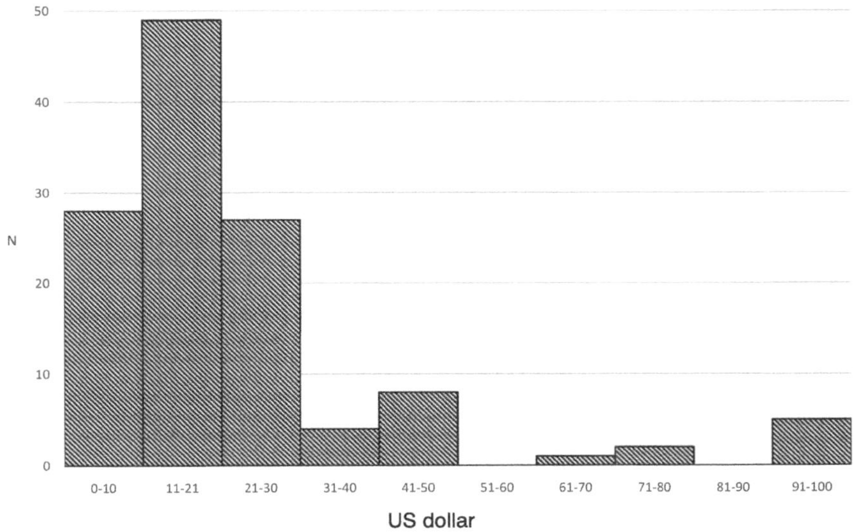

Figure 15.1 Frequency distribution of respondents' willingness to pay in USD (mean = 23.35 USD); the 0–10 category includes those who were unwilling to pay the current entrance fee (respondents did not give a monetary value with this response).

Variables found to be statistically related to willingness to pay in similar CVM studies include age, gender, income, education, nationality, number of people in the group, type of travel group, travel experience, visitors' satisfaction and employment status, among others (Brown and Henry 1989; Navrud and Mungatana 1994; Israngkura 1997; Day 2000; Walpole et al. 2001; Mmopelwa et al. 2007; Shultz et al. 1998; Baral et al. 2008; Lee and Han 2002; Nepal 2007). With the exception of age, none of these variables included in the questionnaire was found to be statistically related to respondents' willingness to pay. Those travelling independently, as opposed to in a tour group, rated their experience at the park more highly. Those who had not participated in a trekking trip prior to visiting Chitwan National Park also tended to provide a (marginally) significantly higher rating for their experience in the park. In neither case was this reflected in their willingness to pay 'bids'. From the point of view of their expressed willingness to pay, tourists visiting the park appear remarkably undifferentiated.

A variety of reasons were given to support the respondents' willingness to pay 'bids'. These are grouped into common themes in Table 15.2 (reasons why respondents were willing to pay an increased entry fee) and Table 15.3 (reasons why respondents were not willing to pay an increased entrance fee).

Many of the positive comments were encouraging as respondents thought they should pay a higher price because the quality of their experience at Chitwan National Park was very positive and conservation of the park was seen as important. In the Annapurna Conservation Area in the Nepali Himalaya, Baral and

Table 15.2 Reasons why respondents were willing to pay an increased entry fee (proportions based on all reasons expressed, positive or negative)

Conditional support, only if funds were judiciously used: 'I would be happy to pay more knowing that it was going towards the park and wildlife conservation' 'I would pay more if I knew where the money was going'	19.4%
Conditional support, only if re-entry was possible: 'A higher fee is fine, but re-entry/multi-day pass is necessary'	10.4%
The entrance fee is too low; if it were higher it could provide more money to improve management and conservation	7.8%
The price is low / paying more is fine	6.5%
Conditional support, only if tourism was better managed: 'I would pay more if there was better tourist management' 'There needs to be a positive change to raise prices'	6.5%
The experience at Chitwan National Park is worth a higher price	3.9%
'Excellent value here'	3.9%
Entrance fees should be more similar to those of other big national parks: '2000 NPR like at Annapurna'	3.9%
'I came to see the park so would pay whatever the asking entry fee'	2.6%
'I expect costs to increase over time'	1.3%

Table 15.3 Reasons why respondents were not willing to pay an increased entrance fee (proportions based on all reasons expressed, positive or negative)

You get what you pay for, which is not very much; disappointed; experience is primitive	10.4%
Show the tourists where the money goes, like at Annapurna: 'I have no idea what entrance fees are spent on' 'More information is needed'	7.8%
The price is perfect, should not be higher	6.5%
Mass tourism makes the area undesirable: 'No tourism organization here' 'They are selling ecotourism, but is this sustainable?'	3.3%
The price is too high already; cannot pay more	2.6%
Fees do not go to the park; higher price would not help	1.3%
Price difference between Nepali and Foreign tourists is not fair	1.3%
On a group tour – no interest in or knowledge of money and activity prices	1.1%

colleagues (2008) conducted a contingent valuation study of trekkers and found that most would be willing to pay an entry fee considerably higher than the then current fee of 2,000 NPR (approximately 27 USD) – the most common reason given was a desire to better protect the environment. In the same area, Nepal (2007) found that trekkers expressed a willingness to pay an additional 'eco-fee' averaging 25 per cent of the entry fee. These observations are generally consistent with the results from Chitwan National Park.

However, three of the ten most common reasons respondents were willing to pay more were based on conditional support. The most common condition was simply that respondents sought to know where the money was going. This is consistent with the literature: both Walpole et al. (2001) and Baral et al. (2008) found that visitors were willing to pay higher entrance fees if information is made available on how the revenues were being used for conservation and upkeep of the park.

There was a sense conveyed by the large majority (88 per cent) of respondents that they did not know where the entrance fees were being directed, and had they known, they might have been willing to contribute much more. This is generally consistent with findings in Thomson's (2007) research in the Annapurna Conservation Area in the Himalayan region of Nepal. Chitwan National Park has many successful community development programmes in the local area, supported by revenue from the park's entrance fees (Lehmkuhl 2000; Mueller-Boeker 2000; Hjortso et al. 2006; Jones 2007; Spiteri and Nepal 2008b). Had information on these programmes been available, visitors likely would have had an even more positive image of the park.

In conversation with the respondents they commonly made reference to the numbers of tourists and to litter, waste disposal and pollution problems. They also noted the lack of evidence of tourist development plans and organization in the area. An increasing proportion of tourists visiting Chitwan National Park may be at least somewhat disappointed with their experience if the number of visitors to the area continues to rise.

Conclusions and recommendations

Chitwan National Park is considered to be the best surviving example of the natural ecosystems of the lowland Terai region; it was Nepal's first national park and has been designated as a UNESCO World Heritage Site. It is also the most visited national park in the nation. Those involved in the park's management face challenges of protection, conservation, addressing the concerns of those living adjacent to the park, and managing increasing numbers of tourists.

Foreign tourists contribute 90 per cent of park entrance fee revenues. This study indicates that the economic value which tourists hold for Chitwan National Park is not being fully captured with the current entrance fee structure. The evidence suggests that foreign tourists are willing to pay more for entry into the park. This would particularly be true if combined with a communication programme to explain the uses to which the park's revenues are put. An increase in the entry fee could provide valuable additional resources which could go towards conservation efforts in the park itself and supporting sustainable development initiatives in buffer zone communities.

Any possible decrease in visitation as a result of an increased entry price might not be seen as negative: a decrease may help to curb some of the undesirable impacts of over-visitation (UNESCO 2003; WWF Nepal 2006; DNPWC 2008) such as litter, pollution, overuse of resources, wildlife disturbance and conflicts, and other forms of natural resource damage. A confounding issue is the effect of the numbers of domestic and SAARC tourists, who represent 32.5 per cent and

11.8 per cent of visitors respectively (Table 15.1), and the effect of their low entry pricing. Further study is warranted.

References

Ahmed, S.U. and Gotoh, K. (2006) *Cost-Benefit Analysis of Environmental Goods by Applying the Contingent Valuation Method: Some Japanese Case Studies*, New York: Springer.

Arrow, K.J., Solow, R.M., Portney, P.R., Leamer, E.E., Radner, R. and Schuman, H. (1993) 'Report of the NOAA panel on contingent valuation: report to the National Oceanic and Atmospheric Administration', *Federal Register*, 58 (10): 601–14.

Aylward, B. and Barbier, E.B. (1992) 'Valuing environmental functions in developing countries', *Biodiversity and Conservation*, 1 (1): 34–50.

Baral, N., Stern, M.J. and Bhattarai, R. (2008) 'Contingent valuation of ecotourism in Annapurna Conservation Area, Nepal: implications for sustainable park finance and local development', *Ecological Economics*, 66 (2/3): 218–27.

Barbier, E.B., Acreman, M. and Knowler, D. (1997) *Economic Valuation of Wetlands: A Guide for Policy-Makers and Planners*, Cambridge, UK: IUCN.

Bateman, I., Langford, I. and Rasbash, J. (1999) 'Willingness-to-pay question format in contingent valuation studies', in I. Bateman and K. Willis (eds) *Valuing Environmental Preferences: Theory and Practice of the Contingent Valuation Method in the US, EU and the Developing Countries*, New York: Oxford University Press.

Bhattarai, K., Conway, D. and Shrestha, N. (2005) 'Tourism, Terrorism and Turmoil in Nepal', *Annals of Tourism Research*, 32 (3): 669–88.

Bhusal, N. (2007) 'Chitwan National Park: a prime destination of ecotourism in central Terai region, Nepal', *The Third Pole: Journal of Geography Education*, 5 (7): 70–5.

Bookbinder, M.P., Dinerstein, E., Rijal, A., Cauley, H. and Rajouria, A. (1998) 'Ecotourism's support of biodiversity conservation', *Conservation Biology*, 12 (6): 1399–1404.

Brockington, D. and Igoe, J. (2006) 'Eviction for conservation: a global overview', *Conservation and Society*, 4 (3): 424–70.

Brown, G. and Henry, W. (1989) 'The economic value of elephants', Paper 89–12, London: London Environmental Economics Centre.

Carson, R., Wright, J., Carson, N., Alberini, A. and Flores, N. (1995) *A Bibliography of Contingent Valuation Studies and Papers*, San Diego, CA: Natural Resource Damage Assessment Inc.

CBS (Central Bureau of Statistics) (2008) *Environmental Statistics of Nepal, 2008*, Kathmandu: Government of Nepal.

CNPPA (Commission on National Parks and Protected Areas) (1995) *Economic Assessment of Protected Areas: Guidelines for Their Assessment*, Gland: IUCN.

Day, B. (2000) *A Recreational Demand Model of Wildlife-Viewing Visits to the Game Reserve of Kwa Zulu Natal Province of South Africa*, London: Center for Social and Economic Research on the Global Environment.

Dixon, J.A. and Sherman, P.B. (1990) *Economics of Protected Areas: A New Look at Benefits and Costs*, Washington, DC: Island Press.

DNPWC (Department of National Parks and Wildlife Conservation) (2008) *Annual Report, Shrawan 2064–Ashad 2065, 2007–2008*, Kathmandu: Ministry of Forests and Soil Conservation.

—— (2010) *Publications*, Kathmandu: Ministry of Forests and Soil Conservation.

Fromm, O. (2000) 'Ecological structure and functions of biodiversity as elements of its total economic value', *Environmental and Resource Economics*, 16 (3): 303–28.

GoN (Government of Nepal) (2008) *Nepal Tourism Statistics 2008: Annual Statistic Report*, Kathmandu: Singha Durbar.

Hanemann, M.W. (1994) 'Valuing the environment through contingent valuation', *Journal of Economic Perspectives*, 8 (4): 19–43.

Hausman, J. (1993) *Contingent Valuation: A Critical Assessment*. Amsterdam: Elsevier Science Publishers.

Hjortso, C.N., Straede, S. and Helles, F. (2006) 'Applying multi-criteria decision-making to protected areas and buffer zone management: a case study in the Royal Chitwan National Park, Nepal', *Journal of Forest Economics*, 12: 91–108.

ICEM (International Center for Environmental Management) (2003) *Economic Valuation: Its Use in Protected Area Management. Lessons Learned from Global Experience. Review of Protected Areas and Development in the Lower Mekong River Region*, Indooroopilly, Queensland, Australia.

Israngkura, A. (1997) *Environmental Valuation: An Entrance Fee System for National Parks in Thailand*, Ottawa: International Development Research Center.

Jones, S. (2007) 'Tigers, trees and Tharu: an analysis of community forestry in the buffer zone of the Royal Chitwan National Park, Nepal', *Geoforum*, 38 (3): 558–75.

Lee, C. and Han, S. (2002) 'Estimating the use and preservation values of national parks' tourism resources using a contingent valuation method', *Tourism Management*, 23: 531–40.

Lehmkuhl, J.F. (2000) 'The organization and human use of Terai riverine grasslands in the Royal Chitwan National Park, Nepal', in C. Richard (ed.) *Proceedings from a Workshop on Grassland Ecology and Management in Protected Areas of Nepal*, Kathmandu: International Center for Integrated Mountain Development.

Luzinda, H. (2008) 'Mobile boundary and mobile people: involuntary resettlement of the Benet people in Mt. Elgon National Park, Uganda', unpublished thesis, Department of International Environment and Development Studies, Norwegian University of Life Sciences.

McLean, J. and Straede, S. (2003) 'Conservation, relocation, and the paradigms of park and people management: a case study of Padampur villages and the Royal Chitwan National Park, Nepal', *Society and Natural Resources*, 16: 509–26.

Mitchell, R. and Carson, R. (1989) *Using Surveys to Value Public Goods: The Contingent Valuation Method*, Washington, DC: Resources for the Future.

Mmopelwa, G., Kgathi, D. and Molefhe, L. (2007) 'Tourists' perceptions and their willingness to pay for park fees: a case study of self-drive tourists and clients for mobile tour operators in Moremi Game Reserve, Botswana', *Tourism Management*, 28: 1044–56.

Mueller-Boeker, U. (2000) 'Ecotourists in Nepal: the example of the Royal Chitwan National Park', in R.P. Thapa and J. Baaden (eds) *Nepal: Myths and Realities*, Delhi: Book Faith India.

Navrud, S. and Mungatana, E.D. (1994) 'Environmental valuation in developing countries: the recreational value of wildlife viewing', *Ecological Economics*, 11: 135–51.

Nepal, S. (2002) 'Involving indigenous peoples in protected area management: comparative perspectives from Nepal, Thailand, and China', *Environmental Management*, 30 (6): 748–63.

—— (2007) 'Ecotourists' importance and satisfaction ratings of accommodation-related amenities', *Anatolia*, 18 (2): 255–76.

Pearce, D. (1992) *Economic Valuation and the Natural World*, London: Centre for Social and Economic Research on the Global Environment.

—— and Moran, D. (1994) *The Economic Value of Biodiversity*, London: Earthscan.

Rietbergen-McCracken, J. and Abaza, H. (2000) *Environmental Valuation: A Worldwide Compendium of Case Studies*, London: Earthscan.

Samdin, Z. (2008) 'Willingness to pay in Taman Negara: a contingent valuation method', *International Journal of Economics and Management*, 2 (1): 81–94.

Shultz, S., Pinazzo, J. and Cifuentes, M. (1998) 'Opportunities and limitations of contingent valuation surveys to determine national park entrance fees: evidence from Costa Rica', *Environment and Development Economics*, 3 (1): 131–49.

Spiteri, A. and Nepal, S.K. (2008a) 'Evaluating local benefits from conservation in Nepal's Annapurna Conservation Area', *Environmental Management*, 42 (3): 391–401.

—— (2008b) 'Distributing conservation incentives in the buffer zone of Chitwan National Park, Nepal', *Environmental Conservation*, 35 (1): 76–86.

Straede, S. and Treue, T. (2006) 'Beyond buffer zone protection: a comparative study of park and buffer zone products' importance to villagers living inside Royal Chitwan National Park and to villagers living in its buffer zone', *Journal of Environmental Management*, 78 (3): 251–67.

TAL-WWF (Terai Arc Landscape Project) (2010) 'Terai Arc Landscape Project'. Online, www.wwfnepal.org/our_solutions/conservation_nepal/tal/project/ (accessed 10 March 2011).

Thomson, A. (2007) 'The Annapurna Conservation Area Project: tourists as agents of development and environmental management in the high Himalaya?' *International Journal of Environment and Sustainable Development*, 6 (4): 405–21.

Tietenberg, T. (1996) *Environmental and Natural Economics*, New York: Harper Collins College Publishers.

UNESCO (2003) 'Initial management effectiveness evaluation report: Royal Chitwan National Park, Nepal, August 2003'. *Enhancing Our Heritage Project: Monitoring and Managing for Success in Natural World Heritage Sites*. Online, http://whc.UNESCO.org/uploads/activities/documents/activity-331-11.pdf (accessed 10 March 2011).

Walpole, M., Goodwin, H. and Ward, K. (2001) 'Pricing policy in protected areas: lessons from Komodo National Park, Indonesia', *Conservation Biology*, 15 (1): 18–227.

Walsh, R., Lommis, J. and Gillman, R. (1984) 'Valuing option, existence and bequest demands for wilderness', *Land Economics*, 860 (1): 14–29.

Weladji, R. and Tchamba, M. (2003) 'Conflict between people and protected areas within the Benoue Wildlife Conservation Area, North Cameroon', *Oryx*, 37 (1): 72–9.

Wells, M.P. and Sharma, U.R. (1998) 'Socio-economic and political aspects of biodiversity conservation in Nepal', *International Journal of Social Economics*, 25 (2/3/4): 226–43.

Whelpton, J. (2005) *A History of Nepal*, New York: Cambridge University Press.

Whittington, D., Briscoe, J., Mu, X. and Barron, W. (1990) 'Estimating the willingness to pay for water services in developing countries: a case study of the use of contingent valuation surveys in southern Haiti', *Economic Development and Cultural Change*, 38: 293–311.

WWF Nepal (World Wildlife Fund) (2006) *Management Effectiveness Assessment of Protected Areas Using WWF's RAPPAM Methodology*, Kathmandu: Nepal Programme, WWF International.

16 The impact of tourism on Latin American World Heritage towns

Alfredo Conti

The origin of tourism in the modern world is directly linked to the construction of the concept of cultural heritage. From the 1980s onwards, World Heritage properties, considered the most important testimonies of monuments, groups of buildings and sites throughout mankind's history, have become main targets for tourism, especially historic towns and centres that express, sometimes more than other heritage categories, the features of a specific culture. Historic towns and centres constitute the main contribution, from a quantitative point of view, of Latin America and the Caribbean to the World Heritage List. These properties illustrate especially the colonial period, between the sixteenth and early nineteenth centuries, even if a few examples from the late nineteenth and twentieth century are also inscribed on the List. One of the main impacts of inscription on the World Heritage List is an increase in the number of visitors. Although tourism constitutes an opportunity for development, through financial benefits, the creation of jobs or the improvement of public spaces and urban infrastructure, threats to the values of the sites are also frequent, especially in those cases where no adequate planning exists. The conflict between the historic town and the tourism town becomes evident in several aspects: replacement of the traditional population in favour of new building owners, disappearance of traditional functions in favour of new services for visitors, unequal distribution of benefits and impact on traditional ways of life of local communities, who are sometimes subject to visitors' expectations. Even if tourism produces benefits, the lack of proper planning oriented to sustainable tourism causes several kinds of problems regarding authenticity and preservation of the outstanding universal values of these unique or exceptional towns. This chapter presents some common situations and explores the impacts on some specific cases of Latin American towns inscribed on the World Heritage List.

The relationship between cultural heritage and tourism

The origin of tourism in the modern world is closely linked to the recognition and appreciation of cultural heritage. In the eighteenth century, the so-called 'Grand Tour', from which comes the word *tourism*, consisted in trips of intellectuals and artists from different countries of Europe to the Italian peninsula, and especially

Rome, to make direct contact with relics of classical antiquity. The ancient world, especially Roman civilization, had been 'rediscovered' by the culture of the Renaissance, and Rome had become a destination for those who wanted to know and appreciate classic art, which was taken as a source of inspiration for artistic production of the time (Choay 1992). This meant the implementation of infrastructure and equipment to meet the needs and requirements of the travellers, including transportation systems and accommodation, all of which constitutes the background to modern tourist facilities. This initial form of tourism included some components of its current definition: the idea of 'tour' meant that travellers returned to their places of residence once their expectations had been met; they were motivated by cultural purposes, and funds invested in the destination came from their home countries. In short, it was, in modern terms, a practice of cultural tourism, reserved for a select social group in terms of education and economic position.

The major economic, social and cultural changes brought about by industrialization all over the world led to new modalities of tourism; among the rights recognized for workers appeared, especially throughout the twentieth century, free time, holidays and leisure. New social groups were gradually incorporated into the practice of tourism. At the same time, new heritage items expanded the realm of tourism attractions, something that has occurred up to the present time. The interest in nature typical of the scientific field between the seventeenth and eighteenth centuries was transferred to the realm of art, especially by nineteenth-century Romanticism; the natural environment was considered as a source of contemplation and relaxation for body and spirit. The translation of this expansion of the concept of heritage impacted the field of tourism; besides sun and beach and cultural tourism, in our time several other modalities appeared, among them ecotourism and adventure, health, business, religious, gastronomic tourism, etc. Over the twentieth century, and particularly the period after the Second World War, progress in transportation and general improvements in revenues facilitated the access of new social groups to the possibility of travel. Social tourism emerged, which ensured the possibility of holidays for workers, and mass tourism, which has continued to expand, making tourism a top economic activity at the international level and, for many countries and regions, a source of economic growth and overall development of communities.

This qualitative and quantitative development of tourism as economic activity coincides with the extension of the concept of cultural heritage in the second half of the twentieth century. The initial idea of 'historic monuments', reserved for the great creations of humankind throughout history or the buildings and sites that had been the scene of memorable events, was changed, as expressed in the Venice Charter (1964), to include testimonies of a particular civilization, a significant development or a historic event, and the concept of monument referring not only to great works but also modest works that have gained cultural significance over time. This expansion of the concept of heritage implied the consideration of modest popular buildings, vernacular architecture, rural or industrial settlements, etc.

In the late twentieth century, new heritage categories were considered; in 1992, the United Nations Educational, Scientific and Cultural Organization (UNESCO) introduced the concept of cultural landscape, consisting of the joint work between man and nature. In 2005, the notion of heritage routes was included in the *Operational Guidelines for the Implementation of the World Heritage Convention*, which implies the consideration of heritage structures based on roads that played a specific role over a considerable period of time and served for cultural exchanges manifested in tangible and intangible heritage components. Moreover, intangible heritage, consisting of literary and musical traditions, social practices, oral history and traditional knowledge, gained significant ground in the theoretical debate and in the field of heritage management, manifested in the adoption by UNESCO of the Convention for the Safeguarding of Intangible Heritage of 2003. In this framework, historic towns and urban areas became one of the most significant heritage assets, since they express, perhaps better than other heritage categories, all the complexity of the human relationship with the environment, merging tangible and intangible heritage components. In this sense, historic towns and urban areas clearly express the idea of heritage as a symbolic reference to the cultural identity of the community (Prats 1997).

At the turn of the twenty-first century, new economic, social and cultural pressures posed new challenges for managing heritage assets. A detailed account of the social, political, economic and cultural conditions of the past two decades is a task beyond the scope and extent of this work, therefore only some aspects will be mentioned with the purpose of better understanding the situation in relation to impacts on heritage and its link with tourism. It is often said that the twentieth century ended between 1989 and 1991 with the fall of the Berlin Wall and the collapse of the Soviet Union. The end of the Cold War began a new period of 'globalization', whereby the different regions of the world are integrated into a single global capitalist market in which competition in pursuit of positioning has a key role. This process was accompanied by significant advances in information technology, notably the diffusion and use of the Internet, with the possibility of a world in permanent communication and connection. Furthermore, tourism underwent a significant expansion; most social groups are no longer satisfied with what is near in terms of distance, but look to expand their experiences in space and in time, both in real and virtual dimensions. As a counterbalance to globalization, the idea of 'local development' emerged, understood as integral growth and improvement, and not only in economic terms, based on the exploitation and utilization of the available resources for a given territory and the community that inhabits it. Against this background of counterpoint between global and local and competition to win markets and audiences, it is not surprising that countries and, in particular, some cities that have traditionally served as sources of generation and sharing of information have begun to look for a better positioning. This aspiration includes large or medium-sized cities, which seek ways to find a place in the global market, in the pages of the press or on the global virtual highway.

One of the key strategies of positioning is to provide cities with new equipment, usually intended for administration, culture, tourism and entertainment. These operations sometimes result in the revitalization of areas that were degraded or under-developed or comprehensive renovation activities; new urban areas are designed to meet the demands of contemporary life. Attracting visitors usually becomes a key strategy, taking into account the economic benefits involved. Heritage properties, especially urban areas or historic centres, are considered an economic resource capable of adapting, through new uses, to the requirements of present society while lending prestige, and act as a main attraction for tourism development. The World Heritage status of some sites implies a significant added value; inscription on the prestigious UNESCO List becomes a scaling factor, not only in strictly cultural terms but also economically, taking into account that the mere fact of registration acts as an efficient and powerful means of dissemination and promotion.

International governmental and non-governmental organizations involved in the conservation and protection of heritage work hard in seeking a balanced relationship between conservation of cultural and natural heritage and tourism use. Several international documents account for such concern. The United Nations Conference on International Travel and Tourism, held in Rome in 1963, stressed the importance of cultural and natural heritage for tourism, calling for conservation measures. In Latin America, the Norms of Quito (Organization of American States 1967), the final document of a meeting sponsored by the Organization of American States, established in 1967 that monuments should be considered as an economic resource, comparable to natural resources, and that their tourism use can be considered a source of development. The document warned, however, about the need for proper planning to ensure the correct tourism use of heritage, including contribution to conservation.

The International Council on Monuments and Sites (ICOMOS) adopted in 1976 an International Charter on Cultural Tourism. The extraordinary growth of tourism over the last decades of the twentieth century led ICOMOS to review the Charter; a new text was adopted in 1999 (ICOMOS 1999), more suitable to the demands of the moment. This text introduces concepts and recommendations related to the proper interpretation and transmission of heritage values and meanings, with the need to consider the tourism use of heritage as a tool for the comprehensive development of host communities through the idea of participatory planning involving all stakeholders.

World Heritage towns

As defined in Annex 3 of the *Operational Guidelines for the implementation of the World Heritage Convention* (UNESCO 2008), historic towns and urban areas constitute, in quantitative terms, the most significant contribution of Latin America and the Caribbean to the World Heritage List. Some 40 per cent of the cultural sites inscribed on the List and located in this region belong to this heritage category, which represents, at the same time, some 31 per cent of the total number of sites inscribed. Most World Heritage historic towns or urban areas belong to the colonial period (fifteenth to early nineteenth centuries), in some cases with a pre-Hispanic background, such as Mexico City or Cusco; Brasilia constitutes an exceptional case of a World Heritage city of the twentieth century. If we consider that archaeological sites comprise the next biggest category, in quantitative terms, and that most of them correspond to pre-European towns in the Americas, we can conclude that the urban history of the continent is quite well represented on the List, much better than other heritage categories, such as cultural landscapes or monuments.

The inscription of historic towns or urban areas on the World Heritage List has an impact that can be defined and measured in terms of different aspects, with some positive and negative effects for their proper conservation in the framework of an integrated approach that takes into account not only economic aspects but also social ones, in particular those related to the quality of life of local populations. It is well known that World Heritage Sites are usually subject to economic and social pressure, for which the theoretical and operational instruments do not always appear to be adequate. In this sense, the debate started in 2005, with the introduction of the concept of 'historic urban landscapes', tends to update approaches and methodologies to deal with these sites.

Latin American World Heritage towns present common features when compared with others located in different geo-cultural regions but, at the same time, some specificities related to their history, morphology, architectural components and social and symbolic characteristics. At the same time, some particular pressures and requirements are evident. The construction of large cultural, administrative or entertainment complexes, currently common in several World Heritage towns all over the world, does not appear, for the time being, as a real threat in Latin America. On the other hand, the strong impact of tourism, social changes, the lack of adequate maintenance of public spaces and buildings and sanitation problems, which are not common problems in other regions, appear frequently in Latin America as a real threat to the integrity and authenticity of the World Heritage towns.

Some World Heritage towns underwent, over the nineteenth and twentieth centuries, a sometimes explosive extension and densification, since they are capital cities or important economic centres; others, in contrast, have had no major modifications to their original size and relations with the natural environment, in some cases on account of economic and demographic stagnation or their isolation with regard to the main productive or exchange circuits. In the first case, historic centres underwent a process that was in a sense similar to that of European cities: the original nucleus preserved their morphologic features and townscape and their function of institutional core of the city, and new urban areas absorbed the functional complexity implied by modernization. These historic centres underwent, in general, processes of substitution of population, becoming sometimes areas that lacked adequate infrastructure, offering bad living conditions and generating social marginalization and functional and physical degradation.

Positive impact of tourism

It is possible to define some indicators of what could be considered the positive impact of tourism on the heritage sites and on the local community. The general idea is that tourism constitutes an opportunity for development, and particular indicators are:

- economic benefits produced by tourism;
- creation of jobs related to tourism;
- improvement of infrastructure and public spaces;
- opportunities for education and training.

With regard to public spaces and infrastructure, World Heritage towns and urban areas are usually a target for improvement and enhancement. Often public spaces are well preserved and maintained; adequate urban furniture and infrastructure are provided or improved. This contributes to a general amelioration of the quality of life of the local population and enhances the experience of visitors.

Inscription on the World Heritage List and the associated increase in the number of visitors constitute an opportunity to restore historic buildings. New uses are often related to tourism. It is quite common in Latin American towns for old one-family houses, which are quite difficult to continue in their original use, to be converted into accommodation for visitors. This can be considered an opportunity in two senses. On the one hand, historic buildings are restored and given a new use, generally by private investors, and, on the other hand, visitors may live the experience of lodging in typical historic houses, which is a means for deeper contact with the local culture.

Figure 16.1 San Miguel de Allende, Mexico. The improvement of public spaces.

Figure 16.2 Valparaíso, Chile. Historic building transformed into a hotel.

Threats caused by tourism

Inscription on the World Heritage List and the tourism use of a town implies undoubtedly some opportunities for local development. But threats usually emerge, especially if no adequate planning has been done and implemented. In this sense, we can notice some differences between Latin American towns and those belonging to other regions of the world: since tourism is considered primarily a source of financial revenues, policies focus on promoting economic benefit and there is no balance with social or environmental aspects. Some of the threats to be discussed in detail are replacement of the traditional population, gentrification, threats to the authenticity of the sites and the impact on traditional ways of life.

Replacement of the traditional population, gentrification

A frequent aspect related to the management of historic urban areas and their tourism use is the displacing of the traditional population or the depopulation of historic neighbourhoods. This phenomenon is strongly linked to a process that can be noted equally in developed and underdeveloped countries. The inscription of an urban area on the World Heritage List implies generally the improvement of public spaces, services and infrastructure. This leads to a rise in the market value of urban land and buildings and so almost inevitably to the replacement of the population. Traditional inhabitants of many historic areas are sometimes displaced by self-decision since they prefer to sell their houses and buy new ones outside heritage areas, something that produces financial benefit. In other cases, they are 'pushed' by the pressure of investors or by groups with major revenues that desire to buy properties in the prestigious historic areas.

The phenomenon of gentrification means the replacement of the typical population of a given urban neighbourhood with another population of greater financial resources or more exalted social position. The truth is that this process takes place in urban centres around the world, though with higher incidence in economically disadvantaged countries. The reason for this process lies primarily in the fact that the buildings are purchased by individuals or entities, usually turning them to different uses from those of the original. Thus, it is common to find old homes converted into hotels, restaurants or shops or even maintaining a residential use but occupied as secondary homes for short periods throughout the year. This means that the neighbourhoods gradually lose their population, which leads at the same time to a crisis regarding some aspects of authenticity. In this case, although the tangible components of the buildings may be in a good condition, or even improved with respect to their previous state, a loss of authenticity of function emerges.

Threats to authenticity

With respect to the alteration or distortion of heritage values and message, it is necessary to consider heritage as a set of tangible assets to which values related to history, art or science are assigned. In this sense, heritage is a carrier of meanings that we try to transmit, through the conservation of the material substance, from one generation to another. The proper understanding and interpretation of such values are therefore essential in order to comprehend the true meaning of heritage, to ensure its proper use and to preserve its authenticity, understood not only as the preservation of the tangible components but also the intangible ones such as functions, vocations, associated traditions, etc. In this sense, a conflict often observed with the spread of mass tourism is that heritage becomes a sort of spectacle and object of consumption, without achieving the adequate transmission and understanding of its values.

It may happen that while a heritage site is well preserved and its carrying capacity or limits of change are maintained at appropriate levels, its dedication to tourism involves risks to its authenticity. This is a situation frequently observed in some heritage categories such as historic centres or urban areas. According to current theory, authenticity is verified in several ways, taking into account tangible and intangible components; thus authenticity includes the consideration of materials, shape and design, setting, functions and vocations, meanings and traditions associated with specific sites.

When we refer to threats to authenticity, we refer not only to damage that can occur to the material components of heritage but also to the risks posed to the intangible aspects that influence authenticity. Thus many historic towns or centres are very well preserved; both buildings and public spaces have good and proper maintenance. But excessive devotion to tourism means that the entire neighbourhood is devoted to visitors; all business is dedicated to the tourist, the old residences are now hotels or devoted to gastronomic uses, etc. The problem then is that while the material substance may be well preserved, the area has lost or drastically changed its meaning and its original functions, so there are aspects of authenticity that are really at risk. This tends to be one of the most complicated issues in the management of heritage sites.

Impact on traditional ways of life

Another type of impact is related to social aspects and, especially, to the relationship between the local community and visitors. Particularly when there is an economic asymmetry between them (more specifically in the case of tourism in economically disadvantaged regions), it is common for residents to consider visitors as an 'opportunity' to secure or increase their income. This can include the selling of typical products or the adoption of behaviours that are expected to have an impact on the visitor. It is common to find people who dress or act in a way that is not part of their daily lives; they become part of the stereotypical image of the site and, therefore, something that the tourist expects to see. The problem is that in this way the resident community, or some of its members, are at the service of visitors' expectations; this is another manifestation of the threat to authenticity, in this case lifestyles, habits, behaviours, etc.

Intangible heritage is fundamental not only to determining the outstanding universal value but also to the authenticity of World Heritage properties. In spite of problems of depopulation and gentrification, most Latin American World Heritage towns retain a very rich intangible heritage made up of, among other components, music, gastronomy and traditions. This intangible heritage is also a tourist attraction and could be jeopardized if adequate safeguarding measures are not defined and implemented. The question is what the limit is to safeguard traditional ways of life or social practices so that they can be preserved as authentic cultural manifestations and not as performances for visitors. What usually happens is that members of the local population act in the way visitors expect them to. There is a sort of alienation of the local population in the expectations of visitors; this is one of the most important threats to authenticity.

Within the social aspects some 'contradictions' appear between what could be called a positive effect on tourism and reality. When talking about tourism opportunities, we have referred to the economic benefit, to the possibility of improving urban spaces and to the provision of infrastructure and equipment. Often the economic benefit is not evenly distributed among the resident community; although there is a general improvement in the urban space and infrastructure, some sites are inaccessible to the residents. Many times the costs of access to cultural facilities and entertainment are fixed in terms of tourism, making them inaccessible to local people.

Some specific cases

In order to illustrate some threats, a few examples of Latin American World Heritage towns and urban areas will be presented. This selection is not intended as criticism of those cases, but they are good examples of the opportunities and threats of tourism in World Heritage towns.

The first case is Cartagena de Indias, in Colombia, inscribed on the World Heritage List in 1984, one of the most beautiful and well preserved historic centres in Latin America. Some of the effects mentioned above can be clearly noticed in the historic centre, such as the improvement in the general condition of the public spaces and the possibilities of new uses for historic buildings. There are different areas within the historic centre. Those which exhibit the best state of conservation correspond to the Cathedral neighbourhood; the state of conservation of the public spaces and buildings is quite good, but the area has undergone a process of depopulation in favour of dedicating buildings to tourism services. Several historic houses have been restored, and very well restored, but they are occupied by people who use them as a second home. This means that these buildings are occupied only during holiday periods or weekends, being closed the rest of the year.

On the other hand, there are the areas where the pressures of tourism are not so evident as yet, where the traditional local population still lives. In the case of Cartagena, the neighbourhood of Getsemaní is the place of residence of the

Figure 16.3 Cartagena, Colombia. Good state of conservation of public spaces and buildings in the historic centre.

Figure 16.4 Cartagena, Colombia. Getsemaní neighbourhood. Popular housing retaining traditional ways of life.

low-income traditional population. Although the state of conservation of the buildings is not as good as in the Cathedral neighbourhood and there are some problems with infrastructure, we can still see here the traditional ways of life and uses of public space. Authenticity is noticeable with regard not only to tangible heritage components but also to intangible components.

Another interesting example is the city of Colonia del Sacramento in Uruguay, whose historic centre was inscribed on the World Heritage List in 1995. This is an interesting example of the merging of different urban and architectural features, since the village was established by the Portuguese at the end of the seventeenth century and passed definitively to Spain at the end of the eighteenth century. Although there are no impressive architectural monuments, the historic centre retains much of the typical atmosphere of a colonial town, amplified by the setting of the historic centre on a peninsula by the Plata River.

Even before inscription on the World Heritage List, the historic centre was subject to tourism pressure and depopulation. At the beginning of the twenty-first century, 330 inhabitants lived in the historic centre; at the same time, one and a half million visitors arrived in Colonia in 2008. It is worth mentioning that the town is located some two hours by car from Montevideo, the capital city of Uruguay, and 50 minutes by boat from Buenos Aires, the capital city of Argentina, with a population of some 10 million inhabitants and an important international

Figure 16.5 Colonia, Uruguay. Typical colonial atmosphere next to the Plata River.

Figure 16.6 Colonia, Uruguay. Street space used for restaurants and cafés.

tourism destination. Visits to Colonia are usually one day excursions from Buenos Aires, and this has a significant impact on the town.

Public spaces and architectural heritage are very well preserved in Colonia, and the local government makes significant investment in conservation and maintenance. The impact of tourism on the public space is noticeable; some streets have been closed to traffic and have become outdoor cafés or restaurants. Gentrification is also important in this town. Due to the increase in the price of plots and buildings, traditional inhabitants preferred to sell their houses in the historic centre and move to other neighbourhoods, with a significant economic benefit. Houses in the historic centre have generally been bought by foreigners who use them as secondary residences. In other cases, buildings have been devoted to new uses related to tourism, such as shops or diverse types of lodging facilities.

Conclusion

The necessity of fostering sustainable tourism becomes a must in these types of site since they bear outstanding universal values which make them unique or exceptional properties that deserve to be preserved in the full richness of their authenticity. World Heritage towns or historic centres act as significant tourism attractions; it is necessary to define and implement measures that prevent such sites from becoming only a beautiful place for tourism, something that could lead to prioritization of the expectations and desires of visitors to the detriment of local communities.

In the postgraduate course on heritage and sustainable tourism delivered at the UNESCO Chair on Cultural Tourism, Argentina, focus is placed on some aspects that could be mentioned as possible recommendations for action:

- financial benefits equally distributed;
- respect for traditional ways of life;
- respect for the authenticity of tangible and intangible heritage components;
- balance between the needs of visitors and those of local communities.

These could help to achieve a healthy balance between the two aspects that characterize World Heritage towns: they are precious testimonies related to specific geo-cultural contexts but, at the same time, spaces for the everyday life of living communities.

Bibliography

Borié, A. and Denieul, F. (n.d.) *Méthode d'analyse morphologique des tissus urbains traditionnels*, Paris: UNESCO, Etudes et documents sur le patrimoine culturel (3).

Choay, F. (1992) *L'allégorie du patrimoine*, Paris: Seuil.

Conti, A. (2007) 'El paradigma urbano en revisión.' Paper presented at the International Conference on Historic Urban Landscapes in the Americas, Olinda, November 2007.

Conti, A. and Cravero Igarza, S. (2010) 'Patrimonio, comunidad local y turismo: la necesidad de planificación para el desarrollo sostenible,' *Notas en Turismo y Economía*, 1, La Plata: Facultad de Ciencias Económicas, Universidad Nacional de La Plata. Online, www.econo.unlp.edu.ar/notastye (accessed 1 June 2010).

ICOMOS (1993) *Tourism at World Heritage Cultural Sites: The Site Manager's Handbook*, Colombo: ICOMOS.

—— (1994) Nara Document on Authenticity. Online, www.icomos.org (accessed 15 October 2008).

—— (1999) International Charter on Cultural Tourism. Online, www.icomos.org (accessed 15 October 2008).

ICOMOS and Grupo de trabajo de países Iberoamericanos (2007) 'Foro sobre ciudades históricas y paisaje histórico urbano.' Unpublished paper.

Organization of American States (1967) Norms of Quito. Online, www.icomos.org/ (accessed 20 May 2009).

Organization of World Heritage Cities (2005) 'Conclusions of the 8th Congress of OWHC, Cuzco.' Unpublished document.

Prats, L. (1997) *Antropología y patrimonio*, Barcelona: Ariel.

Treserras Juan. J. (2005) 'El patrimonio como generador de desarrollo a partir del turismo.' Online, www.fundacioabertis.org/rcs_jor (accessed 20 May 2009).

UNESCO, World Heritage Centre (2007) 'Information Document by the World Heritage Centre on the Development of a Revised UNESCO Recommendation on the Conservation of Historic Urban Landscapes.' Working paper, unpublished.

UNESCO, World Heritage Committee (2008) *Operational Guidelines for the Implementation of the World Heritage Convention.* Online, http://whc.UNESCO.org/ (accessed 20 May 2009).

Van Hoff, H. (2004) *El estado del Patrimonio Mundial en América y el Caribe. Informe Periódico 2004*, Montevideo: Oficina UNESCO Montevideo.

17 Visitor management in sensitive historic landscapes

Strategies to avoid conflict in Hadrian's Wall World Heritage Site

David McGlade

Hadrian's Wall World Heritage Site is one of the UK's iconic historic landscapes; it is also one of its most protected and is endowed with a stable of archaeological, nature conservation and landscape designations. 23 May 2003 witnessed the opening of Hadrian's Wall Path National Trail, a 135 kilometre recreational walking route that shadows the historic line of its eponymous Wall. By the mid-1970s it was apparent that the archaeological fabric of the Wall's central section was suffering from erosion and the initial driver behind the development of the Trail was not the development of tourism but rather a focus on its conservation and protection. That has largely been achieved and today, despite the overall increase in the number of people walking the course of the Wall, the Trail and monument are in a better condition than at any time in their recent modern history.

The Trail project is acknowledged as a case study in countryside and heritage management. It has been achieved by adherence to four underpinning aims: to provide the public with a quality walking amenity; to protect the World Heritage Site's archaeological and nature conservation interests; to maximise the economic gains from walking-based tourism for the benefit of the host community; and, crucially, to minimise any inconvenience that visitors might cause to farmers and other local people.

The Trail is still in its early years but already it has influenced the appearance and character of the World Heritage Site and, in pioneering novel approaches to visitor management, has managed to achieve considerable success. A rigorous field-based monitoring regime ensures that management decisions are based on reliable data that in turn supports the Trail's long-term aims and objectives. However, global economic uncertainty and government reforms, at least in the short term, may yet place strains on the Trail project's ability to deliver a successful conservation management regime. The challenge is to ensure that caring for our heritage does not become a thing of the past.

The recent modern history of Hadrian's Wall

Hadrian's Wall World Heritage Site is today a case study in heritage access management; its visitors are more effectively managed and the archaeological resource better protected than at any time in the past 60 years. This was not always the case. In 1976 a catalogue of visitor-oriented problems was documented in a critical report (Dartington Amenity Research Trust 1976) which highlighted visitor overcrowding and erosion to the fabric of the monument as symptomatic of a complete lack of strategic thinking towards any kind of coordinated management.

The year 1976 witnessed record visitor numbers to the Wall, which lacked the means to ensure its survival and the outlook was bleak. Dartington delivered a timely warning that the Wall was on borrowed time and that without a change of direction the archaeological record, the very thing that people come to visit, would be lost without trace. The Wall's predicament by the mid-1970s, however, was by no means unique. By this time the UK's suite of landscape designations was well established; its National Parks, Sites of Special Scientific Interest and other protected landscapes, including the provision of public access to the countryside, were the product of well-intentioned post-Second-World-War legislation. At the same time increased car ownership meant improved personal mobility, and for some rural destinations a significant increase in visitor numbers resulted in their being dubbed 'honey-pot' sites, especially those suffering from environmental degradation. Many destinations derived significant economic advantages but the benefits were very localised and not shared over a very wide area.

Dartington focused attention on the Wall's 'honey-pot' sites, which became a case study in their own right into the problems associated with sensitive historic sites. By the 1970s some of the issues were long-standing and the integrity of the monument itself was at risk; the archaeological record, in some cases, was literally tumbling down the slope. The new countryside management profession was still in its infancy and lacked the ability to translate landscape policy into site-based solutions and Hadrian's Wall, in common with many UK heritage sites, would end up playing catch-up.

In 1984 the Hadrian's Wall Consultative Committee published *The Strategy for Hadrian's Wall* (Hadrian's Wall Consultative Committee 1984). The Strategy and its principal recommendations – to establish a long-distance National Trail the length of the Wall, and also to manage the landscape as an access corridor – would result in the modern access management systems employed today in the World Heritage Site. Its aims were straightforward: to spread the visitor load along the entire length of Hadrian's Wall; to develop a corridor of circular routes within the Wall's immediate corridor; to promote measures to ensure the conservation of the monuments; and to promote better understanding and awareness of the Wall and its environs.

The Strategy's clear analysis was insightful and far reaching. The focus today extends beyond the inevitable linearity of the Trail and Wall and embraces a 30-kilometre-wide corridor marketed by Hadrian's Wall Heritage Limited as 'Hadrian's Wall Country'. This landscape-wide approach, novel for 1984, has stood the test of time and survives today as a policy aim in the World Heritage Site's Management Plan prepared by Hadrian's Wall Heritage Limited (2009).

A proposed Hadrian's Wall Path National Trail

The years following the publication of *The Strategy* saw the then adviser to the UK government on countryside issues, the Countryside Commission, set about researching a potential route that was considered to be both archaeologically and agriculturally acceptable and, in 1993, the proposed route was presented to the government (Countryside Commission 1993). The government was asked to evaluate the public benefits, including the potential for economic development, against the risk of further erosion and damage to the Wall and its associated earthworks. The proposal was not universally popular with either the Wall's farmers or the wider archaeological community and, while some questioned the wisdom of actually increasing public access to a monument that was already in places very seriously damaged, after almost twelve months of deliberation the government finally gave its consent.

That consent, however, was not unconditional. English Heritage, the government agency responsible for the protection of the historic landscape in England, insisted that the *Submission* document include a critically important clause, subsequently endorsed in the ministerial consent letter. The clause committed the National Trail project to an undertaking to manage, with very few exceptions, the path surface as a green sward (grass) path. The idea is very simple: the maintenance of an unbroken green sward is considered to be the best way of protecting below ground archaeology, much of it near-surface, and it also protects the monument's setting. The aim would at first sight appear to satisfy both farmers, who want to protect their grazing land, and archaeologists, who want to ensure the protection of the archaeology, although differences of opinion can and do occur. That is the nature of countryside management.

Another tension, although not unique to Hadrian's Wall,and as yet not entirely resolved, is the occasional clash of personality between two strands of the conservation movement. The World Heritage Site's landscape is legally endowed with both archaeological and nature conservation designations and, just occasionally, the management objectives of one will come into conflict with those of the other. In the event of an irreconcilable conflict, therefore, and after exhausting every possible solution, which one should take precedence? Notwithstanding the sometimes robust discussions, it is not an exaggeration to say that the green sward aim has underpinned and influenced every strategic and day-to-day decision taken by the Trail's implementation and management project since its inception in 1995.

A strategy to promote sustainable management

The National Trail, by default, finds itself superimposed on archaeology, part extant, part buried. No other access project in Europe has set itself the challenge of sustainably managing public access to almost 100 km of linear monument, much of it an archaeological earthwork.

Not only is the World Heritage Site actively promoted as a visitor destination, but the region also experiences historically high levels of precipitation where, in a normal year, the underlying soils are at field capacity, or saturated, between November and April, during which time there is a greater risk of erosion. The fact that legislation allows for unrestricted public access for 365 days of the year means that, unless novel solutions are constantly researched, tested and trialled, the Trail and its underlying archaeology arbe at risk of permanent damage, especially during the winter months when there is an increased risk of erosion. Its carrying capacity declines considerably when its soils are saturated, which means that the World Heritage Site's marketing and promotion, the seasonality of visit and the destinations that people are directed to when they arrive, all need to be very carefully considered if they are to support the wider conservation aims. It is impossible to separate the interests of the Wall and the Trail; they are managed today as one and the same.

The Trail's management comprises the following elements:

- clear, long-term aims and objectives;
- an aspiration to achieve synergy of aims with key partner organisations and individuals;
- regular proactive grassland management;
- the continued evolution of a generic Scheduled Monument Consent for practical works undertaken in the field;
- systematic monitoring of the ground/monument conditions (establishing, on a field-by-field basis, the link between soil moisture, people counts and ground conditions);
- policies to influence the seasonality of visit, also visitor behaviour;
- the management of the World Heritage Site as an access corridor (Hadrian's Wall Country) with the development of a suite of other circular walking and cycling routes;
- the use of local volunteers.

Despite the short-term nature of government funding, since 1995 the Trail project has adhered to its clear long-term aim of managing the Trail and monument as a green sward and, together with a number of other important decisions that would prove to be influential, this has given it stability. The appointment in 1996 of independent archaeological advisers (Oxford Archaeology North) was welcomed by the wider archaeological community and helped to allay the concerns of some who were worried that recreational countryside managers would make decisions in an arena with which they were unfamiliar. The Trail, however, has managed uniquely to bring archaeology and countryside management closer together, and a long-term research project into best-practice techniques, materials and ideas for monument management has proved invaluable to both sides.

Proactive grassland management and the protection of near-surface archaeology

The National Trail opened in May 2003 without maintenance funding in place and its general condition, and that of the monument, very soon deteriorated. By the summer of 2004 the grass sward was under severe stress, with several new pinch points and wear lines, due almost entirely to recreational pressure, having appeared. This was unwelcome news and the Trail was subjected to intense criticism and scrutiny in both the local and national media for its apparent failure to deliver the green sward undertaking. The Countryside Agency responded by funding the appointment of the two field staff (lengthsmen) whose daily proactive grassland management since then has resulted in today's generally very satisfactory situation. Almost all of the high priority sites identified in 2004 have been restored and the practical management techniques, formalised by the Trail's generic Scheduled Monument Consent (see below) continue to evolve.

The many archaeological evaluations and watching briefs undertaken when items of Trail furniture (gates, footbridges etc.) are to be installed within the Scheduled Monument area have revealed that the foundations of Hadrian's Wall and other structures and artefacts are invariably near-surface archaeology, sometimes within the top soil horizon, in other words so close to the surface as to be at risk of becoming lost if damage were to occur to the sward. This only serves to underline the importance of the green sward, and the effectiveness of the proactive grassland management, to archaeological conservation. Some 100 or so kilometres of Hadrian's Wall Path, away from the urban areas of Newcastle and Carlisle, are farmed and under grass. It is impossible to patrol and inspect all of it on a daily basis, and problems can and do occur without warning. Rainfall, soil moisture, drainage, angle of slope, people counts, not to mention land use, all contrive on occasion to throw up unexpected wear lines and pinch points. When this happens, it is vitally important that the lengthsmen react with an appropriate management response.

The Trail project has pioneered a novel approach to the landscape-wide management of protected archaeological areas. A generic Scheduled Monument Consent means that statutory permissions for a menu of field-based management prescriptions are pre-approved. This reduces bureaucracy because it avoids the need to apply for consent every time a management task is proposed and, more importantly for the staff concerned, it enables effective in-the-field management in real time in the knowledge that what they propose to do, having already been pre-approved, is legal. The generic Scheduled Monument Consent has also influenced the national agenda, with the introduction of area-wide management agreements featuring in the recent UK Heritage Protection Review. Both the generic Scheduled Monument Consent and the lengthsmen are fundamentally important to the long-term green sward aim. Surface-management techniques are continually researched and tested on a field-by-field basis; without this approach the archaeological integrity of the World Heritage Site would once again very soon become compromised.

Monitoring the Trail and monument condition

All countryside management projects need an effective means of monitoring and recording what is happening in the field and Hadrian's Wall Path is no exception;

without monitoring data it is impossible to plan and make informed management and maintenance decisions. The Trail's monitoring methodology is often cited today as an exemplar within the UK's heritage and countryside management sectors, with the first fixed-point photographic monitoring sites established as early as 1996. Sites were rolled out incrementally in line with the Trail's development and by 2003 every rural Wall-mile had at least one monitoring site. Today the exercise is repeated at some 90+ sites every April, August and November, with images taken for both east and west-bound views – a sequence of images which capture the ground condition before, during and after the main walking season. It provides the Trail staff at Hadrian's Wall Heritage Limited with an unrivalled archive of path and monument condition monitoring data, although care must be taken with its interpretation in order to avoid arriving at false negatives. Experience has underscored the importance of understanding long-term trends in ground condition, and avoiding conclusions based on short-term observations. The data time-line, however, is now long enough to enable informed management decisions that stand up to scrutiny.

The field-based monitoring is followed by an office-based exercise where the images are scored on a scale of 1 to 5. A score of 1 indicates that the grass surface is in almost pristine condition; 3 indicates some exposed soil and early signs of a pinch point developing; 5 shows that the erosion is so severe that bedrock is exposed; and there are intermediate assessments for sites scored 2 and 4.When expressed graphically, the summary data reveals that since November 2004 the condition of the Trail and monument has shown a steady gradual improvement, an improvement that is entirely coincident with the appointment of the lengthsmen and the regime of regular proactive grassland management.

However, the fixed-point photographic exercise, while capturing the state of the Trail and monument, can only ever be a snapshot in time and has to be supported with other fieldwork-based data, including reports from Trail volunteers. The Trail is field-surveyed every autumn by the Trail manager and lengthsmen, accompanied by the Trail's independent archaeological consultant, an invaluable exercise that gives key staff an end-of-season overview of the condition of the World Heritage Site.

National Trail passport scheme and the importance of soil moisture

Monthly Meteorological Office soil-moisture and precipitation data is fed into a statistical model to monitor the complex relationships that exist between visitor numbers, climate and the fixed-point photographic monitoring condition scores. The long-term mean for both reveals that, in a normal year, the World Heritage Site's soils are in moisture deficit between the months of May and October. Conversely, during the winter and early spring months, between November and April, the soils are at field capacity (saturated); this is when they are least capable of supporting the visitor load and the risk of erosion is greater.

The Trail's seasonal passport scheme encourages visitors to walk the Trail during the (normally) drier spring, summer and autumn months when the underlying soils are in soil moisture deficit and more capable of withstanding wear and tear. The idea is very simple: walkers collect a set of seven passport stamps between 1 May and 31 October each year; only when their card is completed, can they claim their exclusive enamel achievers' badge and certificate. The decision to devise this unique-in-UK-terms marketing instrument was made in 1995 at the outset of the Trail's implementation, when it was realised that the issue of soil moisture and seasonality had hitherto been overlooked. It is very popular with the vast majority of walkers carrying the passport, and there are stories of some, having missed a stamping station during their walk, returning afterwards simply in order to collect a missing stamp. This suggests that the passport scheme influences the decision of some walkers towards the drier months of the year, and it remains firmly fixed as an essential part of the Trail's strategy for sustainable conservation management.

Conservation messages

Visitors are also encouraged to engage with the conservation management process. The Trail project has coined what it calls its *conservation tips*, a couple of very simple measures that can and do make a significant contribution to the conservation of the World Heritage Site. Visitors are asked, when walking in a group, to walk side-by-side instead of single file; by doing so the carrying capacity of the grass sward is effectively doubled without any effort on their part. Small pictorial notices at regular intervals along the Trail help to promote the *conservation tips*, with sufficient numbers of walkers entering into the spirit of the exercise for it to make a noticeable difference to the condition of the grass sward. Although the message may appear counter-intuitive and contra to the message in the UK's mountainous regions where visitors are asked to contain the damage by walking "in" an erosion scar, on Hadrian's Wall the aim is simple – that of preventing erosion from occurring in the first place.

Visitors are also asked, for rather more obvious reasons, to resist the temptation to walk on top of Hadrian's Wall where it survives as extant masonry. Although much of the Wall has been consolidated, walking on it is a potentially damaging activity and original Roman masonry can and does become dislodged. Other more fragile sections of the Wall, known as Clayton Wall, which take their name from John Clayton, a nineteenth-century antiquarian who purchased and rebuilt sections of the Wall, have no structural integrity whatsoever and collapses are not uncommon. However, some visitors, fostered by elements within today's popular journalism, are tempted to re-create the impression of 'walking in the steps of the Romans' atop the Wall. For its part, the National Trail project promotes the World Heritage Site Management Plan's aim of discouraging walking on top of the Wall and has coordinated the design and production of notices with the aim of changing visitor behaviour.

While site-based information is important, it is clearly helpful for visitors to have an awareness of the conservation issues prior to their visit to the World Heritage Site. Not only do the World Heritage Site's main website portals –

www.hadrians-wall.org and www.nationaltrail.co.uk – promote the conservation messages, but every opportunity is also taken to influence other web-based platforms. The project proactively liaises with traditional publishers and print and broadcast media, and has also contributed a case study to Best of Both Worlds,[1] a website aimed at organisers of large countryside events with the goal of promoting responsible behaviour in sensitive natural and historic landscape environments. Publishers of National Trail maps and guides have also responded very positively by publishing the conservation messages, notably Harvey Maps, Footprint Maps, Cicerone, Trail Blazer and Rucksack. This liaison is demanding in terms of staff resources, but the opportunity cost of not developing the relationships would be at the expense of the World Heritage Site.

Conclusion

The Hadrian's Wall Path National Trail project is attempting to do something different. No other countryside heritage/recreation site anywhere in the UK, Europe or beyond provides such extensive public access to an historically unique and sensitive landscape. The risks are potentially great and the watchful eye of the archaeological community is a reminder that scrutiny, and at times criticism, is close at hand. That is perfectly understandable. Archaeology is, of course, a finite resource, and if any part of it is damaged or lost, it can never be replaced; it is as simple as that and when the archaeology in question belongs to a World Heritage Site, then the responsibility is greater still.

Since the outset the boundaries and scope of the Trail's managers have been stretched to the limit and this has led to innovative thinking and solutions to problems. Some of the solutions have at times been met with scepticism and inertia but the project today has an accumulated net experience which enables it to anticipate issues, often before they become apparent, and enact an appropriate management response. The barriers to some of these solutions are not necessarily financial; in fact most of the countryside management undertaken in the field along Hadrian's Wall is relatively low cost and straightforward in terms of its implementation. In order to achieve some of its objectives the project has adapted to the skills and psychology of change management and has demonstrated that once a technique or idea has been shown to work, then it stands a chance of becoming accepted as routine management. For both short and long-term projects the lead organisation has to accept responsibility for relationship management; effective partnerships require a considerable investment in time because individual partners' own aspirations and resources will inevitably change over time.

The evolution of the grassland management approach along the National Trail, as opposed to hard engineering solutions, has had many successful outcomes, in particular where level ground and shallow gradients are concerned. Areas of severe stress recorded in a 1991 condition survey, including exposed pinch points, wear lines and a loss of top soil, today have a healthy grass cover. Wear lines on slopes are more challenging but new ideas have also met with promising results. Today's management techniques and materials, allied to the principle of spreading the visitor load over a marginally wider area, mean that hard landscaping is now very much the option of last resort.

The fixed-point photographic monitoring exercises also demonstrate that the Trail/monument condition has improved significantly since the appointment in 2004 of the lengthsmen. The seasonal passport, one of the Trail's key foundations, also contributes towards this success, and it continues to influence walkers towards what are the normally drier months of the year. For the winter months especially, but available at any time of the year, a suite of alternative walking routes that connect to other more robust destinations helps to spread the visitor load, and also the economic benefits of tourism, over a wider area. The lessons from a very large group of 800 visitors who arrived without warning in January 2003, causing extensive damage to archaeological earthworks, have been learnt. Such large groups are discouraged unless they agree to be managed, and the Trail project today offers specialist advice to event organisers on route design.

It is also now universally accepted that National Trails in England and Wales are the flagships for countryside access management. A growing number of visitors disperse considerable economic benefits to the Trails' host communities, and in recent years many farm-based accommodation and refreshments businesses have opened. Numerous challenges remain, however, and from time to time farmers are unfortunately inconvenienced by walkers, but it is the job of countryside managers to devise solutions to these situations. The difference between 1976 and today is that National Trails exist within the context of a well-managed countryside staffed by experienced countryside management professionals. In the case of Hadrian's Wall Path and the wider World Heritage Site, their management thinking is founded, post-Dartington, on more than 15 years of practical experience in the field. The fact that archaeologists and countryside managers now collaborate towards common objectives is a very positive development. Hadrian's Wall Path is today a largely successful project because of its long-term adherence to clear aims and objectives supported by appropriate resourcing. It is hoped that global economic uncertainty will not challenge that approach.

Note

1 www.bobw.co.uk.

References

Countryside Commission (1993) *The Hadrian's Wall Path Submission to the Secretary of State for the Environment*, Cheltenham: Countryside Commission.

Dartington Amenity Research Trust (1976) *Hadrian's Wall: A Strategy for Conservation and Visitor Services*, Cheltenham: Countryside Commission.

Hadrian's Wall Consultative Committee (1984) *The Strategy for Hadrian's Wall*, Newcastle upon Tyne: Countryside Commission.

Hadrian's Wall Heritage Limited (2009) *The Frontiers of the Roman Empire World Heritage Site 2008 – 14*, Hexham: Hadrian's Wall Heritage Limited.

Index

Note: Page numbers followed by 'f' refer to figures, followed by 'n' refer to notes and followed by 't' refer to tables.